AF248911

Winning the SoC Revolution:
Experiences in Real Design

Winning the SoC Revolution:
Experiences in Real Design

Edited by:

Grant Martin
Cadence Labs
&
Henry Chang
Cadence Labs

 Springer

Library of Congress Cataloging-in-Publication Data

Winning the SoC revolution: experiences in real design / edited by Grant Martin &
Henry Chang.
    p. cm.
   Includes bibliographical references and index.
   ISBN 1-4020-7495-6
   1. Systems on a chip. I. Martin, Grant (Grant Edmund) II. Chang, Henry.

Tk7895.E42W56 2003
004.16—dc21

                        2003051413

Printed in the United States of America.

9 8 7 6 5 4 3 2   SPIN 11330752

springeronline.com

Contents

About the Editors   vii

Acknowledgements   ix

Preface   xi

1. THE HISTORY OF THE SOC REVOLUTION   1
   GRANT MARTIN

2. SOC DESIGN METHODOLOGIES   21
   HENRY CHANG

3. NON-TECHNICAL ISSUES IN SOC DESIGN   47
   RON WILSON

4. THE PHILIPS NEXPERIA DIGITAL VIDEO PLATFORM   67
   J. AUGUSTO DE OLIVEIRA AND HANS VAN ANTWERPEN

5. THE TI OMAP™ PLATFORM APPROACH TO SOC   97
   PETER CUMMING

6. SOC - THE IBM MICROELECTRONICS APPROACH   119
   KATHLEEN MCGRODDY-GOETZ, ROBERT DEVINS,
   MICHAEL D. HALE, MARK KAUTZMAN, G. DAVID ROBERTS,
   AND DWIGHT SULLIVAN

7. PLATFORM FPGAS   141
    PATRICK LYSAGHT

8. SOPC BUILDER: PERFORMANCE BY DESIGN   159
    JESSE KEMPA, SHEAC Y. LIM, CHRIS ROBINSON,
    AND JOEL A. SEELY

9. STAR-IP CENTRIC PLATFORMS FOR SOC   187
    JAY ALPHEY, CHRIS BAXTER, JON CONNELL,
    JOHN GOODENOUGH, ANTONY HARRIS,
    CHRISTOPHER LENNARD, BRUCE MATHEWSON,
    ANDREW NIGHTINGALE, IAN THORNTON,
    AND KATH TOPPING

10. REAL-TIME SYSTEM-ON-A-CHIP EMULATION   229
    KIMMO KUUSILINNA, CHEN CHANG,
    HANS-MARTIN BLUETHGEN, W. RHETT DAVIS,
    BRIAN RICHARDS, BORIVOJE NIKOLIĆ AND
    ROBERT W. BRODERSEN

11. TECHNOLOGY CHALLENGES FOR SOC DESIGN   255
    JOHN M. COHN

Index   297

About the Editors

Grant Martin is a Fellow in the Labs of Cadence Design Systems. He joined Cadence in late 1994. Before that, Grant worked for Burroughs in Scotland for 6 years and Nortel/BNR in Canada for 10 years. He received his Bachelor's and Master's degrees in Mathematics (Combinatorics and Optimisation) from the University of Waterloo, Canada, in 1977 and 1978.

Grant is a co-author of the books *Surviving the SOC Revolution: A Guide to Platform-Based Design*, published by Kluwer Academic Publishers, in November of 1999, and *System Design with SystemC*, published by Kluwer in May of 2002. He co-chaired the VSI Alliance Embedded Systems study group in the summer of 2001. His particular areas of interest include system-level design, System-on-Chip, Platform-Based design, and embedded software.

Dr. Henry Chang received his Sc.B. degree in Electrical Engineering (EE) from Brown University in 1989, his M.S. degree in EE from the University of California at Berkeley (UCB) in 1992, and his Ph.D. in EE on a "Top-Down, Constraint-Driven Design Methodology for Analog Integrated Circuits" from UCB in 1994.

As an Architect at Cadence Design Systems, Inc. from 1995-2002 he has worked on various activities related to system-on-a-chip (SoC). These activities include SoC design methodologies, capabilities, tools, flows, and standards looking at the general SoC problem as well as analog & mixed-signal IP design issues. He chaired the VSI Alliance Mixed-Signal Development Working Group from 1996-2002 and co-chaired the group until 2003. He is one of the co-authors of the book, *Surviving the SOC Revolution: A Guide to Platform-Based Design*. Starting in 2003, he has been a Director in Cadence's Corporate Strategy group developing and driving Cadence's long range business plans.

Acknowledgements

Grant Martin would like to acknowledge, as always, his wife Margaret Steele, and his daughters Jennifer and Fiona. He would also like to acknowledge the encouragement of his father and mother, John (Ted) and Mary Martin.

Henry Chang would like to acknowledge his family - his wife Pora Park, his son Daniel, and his parents Shih-hung & Pan-chin whose help in taking care of Daniel made his contributions to this book possible.

Grant and Henry would also like to acknowledge all of the hard work by the contributors of the chapters in this edited volume. Any mistakes found should be regarded as the responsibility of the editors.

Preface

We are living through a revolution in the design of large complex integrated circuits - the SoC revolution. When they first appeared in the mid-1990's, System-on-Chip (SoC) was arguably just a marketing term. At that point, the semiconductor fabrication process technology had achieved the scales of 350 and 250 nm, allowing the integration of only relatively simple digital systems. But, by the turn of the millennium, with 180, 150 and 130 nm processes available, many design teams were building true SoC devices.

These devices are systems in every sense of the word. They incorporate programmable processors (often with at least one RISC and one DSP), embedded memory, function accelerators implemented in digital logic, complex on-chip communications networks (traditional master-slave buses as well as network-on-a-chip), large amounts of embedded software, both hardware-dependent, middleware and applications, and analogue interfaces to the external world.

But the design theories, methods and tools for designing, integrating and verifying these complex systems have not kept pace with the advanced semiconductor fabrication processes that allow us to build them. The well-known International Technology Roadmap for Semiconductors (ITRS) analyses point to an ever-widening gap between what we are capable of building, and what we are capable of designing. Many techniques have been proposed and are being used to close that gap, including design tool automation and integration from RTL to GDSII, integrated testbenches, IP reuse, platform-based design, and, last but not least, system-level design techniques. These techniques grow in importance as the industry moves to the 90, 65, 45 nm process technology nodes, and beyond.

Although commercial tool providers, large design companies, and academic researchers are doing a good job of developing the theory and practice of many of these techniques, there remains a considerable lack of pragmatic knowledge among practitioners of the leading design methodologies for SoC design. There have been many interesting presentations about IP reuse and SoC at a number of the well-known conferences, and some seminal books have given introductions to many aspects of reuse and SoC, but the real-world perspective of leading SoC design teams has been missing.

In 1999, at the dawn of the SoC revolution, we published a book entitled, *Surviving the SOC Revolution: A Guide to Platform Based Design*. At the time, there were few industrial examples of true SoCs. The content we provided came from intense thinking and a research and development project that began in 1997 to develop methodologies for SoC and platform based design. We are now in the middle of that revolution. Many industrial examples do exist. Now the debate centres on what is the best approach for SoC design, but there is now no doubt that SoCs are here to stay. It is now time to **win** the SoC Revolution. We believe now, in the industry, that there is enough design experience and know-how that we can do so. Therefore, rather than writing the entire book ourselves as we did with *Surviving the SOC Revolution*, we decided the best approach to this book would be as a collected volume authored by the leading practitioners of SoC design.

Readers thus should consider this book to be a sequel to *Surviving the SOC Revolution*, rather than a second edition of it. The two books complement each other: *Surviving the SOC Revolution* focuses in detail on all of the issues involved in SoC design, and gives many possible solutions to designing them using the concept of platform-based design. This book, *Winning the SoC Revolution: Experiences in Real Design*, focuses on know-how, experiences, and solutions. It is indeed possible to move on from mere survival, to mastering SoC.

This book presents a variety of lessons and examples in SoC design, indicating the key areas and most important design methodologies which are considered essential to successfully developing large complex integrated circuits. In Chapter 1, by Grant Martin, we have a brief history of the SoC revolution and SoC development issues, including changing expectations, new design paradigms and the role of important industry standards organizations such as VSIA. Chapter 2, written by Henry Chang, elaborates on the breadth and scope of the design methodologies used for SoC. This chapter also sets the remainder of the book in context, by showing where many of the concepts presented fit into the overall scheme of things. This is then followed by a very important exposition by Ron Wilson which discusses a variety of the organizational, managerial and non-technical design disciplines which are essential to successful SoC design. These considerations are treated from the point of view of design interfaces - between components of design, between suppliers and integrators, between designers, and between the design team and the foundry. Often these areas are forgotten when design teams consider what SoC requires.

Chapters 4-6 contain seminal examples and lessons from some of the leading integrated device manufacturers (IDMs) today, with custom and semi-custom style SoCs. In Chapter 4, a team from Philips talks about the major architectures and design disciplines necessary to develop a successful family of digital multimedia processing SoCs. Chapter 5 from Texas Instruments gives details about the design approaches used with the successful OMAP family of SoCs used in a variety of portable consumer wireless and multimedia processing applications. And finally, in Chapter 6, a contribution from IBM Microelectronics gives strong emphasis to the role of advanced verification methodologies and technologies in SoC development.

In Chapters 7 and 8, we turn from masked fixed-function SoCs to the interesting new developments in reconfigurable logic platforms. Chapter 7 from Xilinx gives an overview of the Xilinx Virtex II PRO platform FPGA, and discusses both design approaches for using these platforms, and possible future reconfigurable platform developments. This is then followed by a chapter from Altera, which discusses their design philosophy and configuration tools for System-on-a-Programmable Chip (SOPC). Reconfigurable logic SoC platforms represent a particularly exciting and intriguing combination of in-field flexibility and programmability, and customized hard cores such as embedded control processors and high-speed IO interfaces.

In Chapter 9, we turn to arguably the most important, or at least one of the most influential, IP providers: ARM. This chapter discusses their move from being a supplier of embedded processor cores and buses, to being a full supplier of processor-centric platforms, with the PrimeXSys platform being discussed as an example of this evolution.

Chapter 10 from Professor Bob Brodersen's team at the Berkeley Wireless Research Centre takes a different tack to the design of complex SoCs. Their somewhat unique design philosophy aims at allowing rapid mapping of complex signal processing applications with embedded control into both a direct silicon realization as an SoC, and onto a rapid prototyping platform for early algorithm and design space exploration and prove-in, and for associated system and SW development.

Finally, in Chapter 11, we are given a strong dose of technology reality. John Cohn of IBM paints the current and future developments in semiconductor processing technology, and the problems and issues raised for SoC design. But his chapter is not just a litany of problems: he provides extensive discussion of how these problems are being tackled at IBM, which as one of the leading SoC design and manufacturing houses, is an excellent example for those wishing to understand advanced process technology and how design tools and methods deal with fundamental issues.

This book is essential reading for all those wishing to learn the secrets of mastering modern SoC design. Although there are many additional important topics which designers must learn and practice, obtaining a basic grasp of the material presented here by the world's leaders in SoC will give the reader a firm grounding in how to start applying the most advanced design principles. If the reader applies the lessons contained within, he or she will have every chance of joining the circle of SoC winners themselves.

Grant Martin
Henry Chang
Berkeley and San Jose
April 2003

Chapter 1

THE HISTORY OF THE SOC REVOLUTION
The Rise and Transformation of IP Reuse

Grant Martin
Fellow, Cadence Berkeley Labs

Abstract: We cover SoC design from its roots in hardware design reuse and the movements in the mid-1990's that led to the creation of a reuse guideline-based approach for block-based design, the formation of the Virtual Socket Interface Alliance (VSIA), and the development of the block-based approach to SoC composition.   We discuss the early successes and failures of IP reuse and what has been learned from these approaches. We will study the most successful SoC design methodology seen to this date: the platform-based design approach. Finally, we conclude by looking ahead to new styles of SoC and new design approaches which may emerge in the future.

Key words: SoC, IP, VSIA, reuse

1.   INTRODUCTION

It has now been many years since the "System-on-Chip" (SoC) revolution started.   This revolution commenced sometime in the mid-1990's, when semiconductor process technology began reaching the point, at 350 and 250 nanometers (or more commonly, 0.35 and 0.25 microns), at which the major processing elements of complete "system" products could begin to be placed on a single die.   For example, second generation cellular phones could, at 350-250 nm, begin to integrate all of the major digital baseband and control elements onto a single silicon substrate, including the RISC control processor, DSP, hardware signal processing elements (e.g. for voice), memory and memory interface, and peripherals.   Arguably, such a digital baseband chip for 2G was far from the complete "system": analogue baseband, RF, analogue power control and terminating resistors were all separate discrete components, albeit all highly integrated onto a small circuit board fitting into the rather large wireless handsets of the day.   But such "systems in silicon" or early SoC clearly delineated the evolutionary direction of consumer systems as semiconductor process technology advanced.   What is a multi-chip integrated chipset today will become a single integrated SoC tomorrow, enabled by more advanced manufacturing processes.   And eventually the problems of integrating analogue, digital, RF and even more exotic structures such as micro-electronic mechanical systems (MEMS), sensors, actuators, lab-on-a chip chemical processing, optical and biological processing elements would all be solved.

What is a System-on-Chip? If we hearken back to our definition from *Surviving the SOC Revolution*, and modify it slightly, we can define an SoC as a complex IC that integrates the major functional elements of a ***complete end-product*** into a single chip or ***chipset***. In general, SoC design incorporates at least one ***programmable processor***, on-chip memory, and accelerating function units implemented in hardware. It also interfaces to ***peripheral*** devices and/or the real world. SoC designs encompass both hardware and ***software*** components. Because SoC designs can interface to the ***real world***, they often incorporate ***analogue*** components, and can, in the future, also include opto/microelectronic mechanical system (***O/MEMS***) components. Figure 1-1 shows an example of such a device.

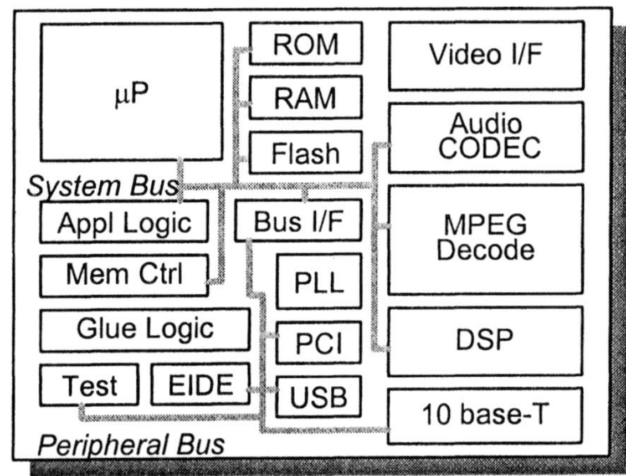

Figure 1-1. A System-on-Chip Device

If we look back in the year 2003 to the optimistic projections of the mid-1990's, all SoC problems seemed daunting but surmountable. If we examine the SoC marketing "hype" of the mid-90's, it seemed that by the turn of the millennium, all problems would be solved and we would be living in a radiant future in which SoCs would be turned out by design teams both rapidly, to meet accelerated time-to-market requirements for advanced products, and with high reliability and tremendous productivity. As we all know, although some elements of this vision have been realised, we are still quite a long way from its full implementation. As in all revolutions, the situation several years after starting is not quite what was predicted during those heady early years. In this chapter we will talk about the history of the SoC revolution, the current status of SoC design, and trends and directions for future SoC evolution - all topics which are elaborated in much more detail in the rest of this book. We will cover both design and electronic design automation (EDA) perspectives, and touch on the issues of design reuse, product development, technology evolution, and their implications.

2.　　THE GLORIOUS HOPE: 1995-1999

In order to trigger revolutionary conditions, discontent must stalk the land, and leaders must arise to capitalise on this discontent. The peasants and proletariat usually suffer under the rigid rule and iron law of an autocratic monarchy or oligarchy, which remains indifferent to their suffering. The design of complex integrated circuits *has* been subject to an "iron law" over the last 40 years - the Iron Law of Moore. Moore's law [1,2], when combined with the incredible progress in semiconductor technology, and the invisible hand of economics, have worked together to crush the spirits of our design proletariat during the 1990's. The well known "design gap" [3] between what the electronics industry is capable of building, and what reasonable design teams are capable of designing in a certain design interval, led to a perceived productivity crisis in IC design by the mid-1990's. This was compounded by tighter time-to-market constraints, and an economic imperative to integrate more and more of the design into a single or small number of circuits in order to achieve the low costs demanded by consumer applications.

There have been many solutions suggested to this design productivity crisis [4,5], but there is no doubt that by the mid-1990's the solution perceived to have the biggest potential impact on design productivity was Intellectual Property (IP - i.e., design IP) reuse. In order to fill the design gap to produce a reasonable sized IC with enough logic and memory to both justify its design and manufacturing cost and deliver enough functionality for that integrated system, well-known models were developed that indicated IP reuse had to move from the low levels prevalent in the mid-90's (0 to 30 percent) to an extraordinarily high level (90 to 99%) within 2 or 3 process generations [6].

Design reuse was felt to be the answer, but it was a little-understood set of *ad hoc* techniques.   In order to achieve the vision of massive design reuse, it was clear that the industry required:

– New design methods to allow for the creation of reusable design (and verification) IP.
– New standards for IP exchange, interfacing and interoperability.
– The emergence of an IP industry to provide the raw material for IP exchange, created with the new design methods, conforming to the new standards, and thus fostering reuse within large companies, between design groups, and via a 3rd party IP industry.
– A cross-industry consensus on what was required to enable reuse.
– Resolution of IP business issues.
– Strong industry leadership to make all this happen.

Three seminal events occurred in the mid-to-late 1990's which both indicated the way forward and led the way during this first hopeful phase of the SoC revolution:

1. The first, and most significant, was the establishment of the Virtual Socket Interface Alliance (VSIA) [7] as the cross-industry organisation that would focus on IP reuse for SoC design - and all of the requirements listed above.   VSIA was founded to help foster the SoC revolution by combining the design skills and knowledge of semiconductor companies, systems companies, and the EDA industry.   After an intensive period of development of its first draft Reuse Architecture document [7], starting in spring of 1996, VSIA was founded and publicly announced in September, 1996.

2. The second event was the creation of the "Reuse Methodology" for "soft" (synthesizable RTL) IP by teams from Mentor Graphics and Synopsys, culminating in the *Reuse Methodology Manual* made available in a controlled way in 1997, and published as a text by Michael Keating of Synopsys and Pierre Bricaud of Mentor Graphics - the *Reuse Methodology Manual* - in 1998 [8].   This seminal text has since gone through two additional editions and has been the first or starting reference on designing reusable IP for a whole new generation of designers.

3. The third event was the creation, in Scotland, of the "Alba" project - a government-industry-academic collective which aimed to transform the Scottish high technology economy into one centred on System Level Integration and IP-based SoC design. This is only one, albeit the leading, earliest and largest example, of a country or regional focus on economic development in high technology by mobilising government, university

and industry resources in a focused collective action for change, via SoC design.

2.1   The Virtual Socket Interface Alliance (VSIA)

The VSIA started informally in the spring of 1996 with various discussions on what it would take to enable better IP reuse. These early discussions involved Toshiba and Cadence Design Systems, followed quickly by Fujitsu and Mentor Graphics. An initial version of the VSIA architecture document, which described IP deliverables for soft and hard IP (firm IP requiring more definition) was created by August, 1996, and in September a formal kickoff meeting and press announcement was held, involving 7 corporate steering group members and a total of 35 VSIA company members in all. This included many major semiconductor, systems and EDA houses.

The first VSIA members meeting was held in October, 1996, at which time there were 50 members, and the opportunity was taken to initiate development working groups (DWGs) in a number of key areas: implementation, manufacturing-related test, analogue/mixed-signal, system level, on-chip bus, and IP protection. The organisation grew rapidly, with over 100 members by March, 1997 and over 160 by November of that year. VSIA peaked in membership in the late 1990's with over 200 members.

In the spring of 1997, VSIA publicly released what it called its Reuse Architecture document [7], outlining a basic set of IP deliverables that made up the notion of a "Virtual Socket" for IP. Given the focus on RTL to GDS II design during the preceding period and the initial formation of VSIA, this was the focus of the basic architecture document. It was expected that additional IP-related deliverables in the domains of test, bus, AMS, IP protection and system-level design would emerge from the development working groups over the next two years.

VSIA also expanded in scope from a primarily North American organisation in its formation, to a worldwide scope, with significant membership in Europe, Japan and Asia, and member centres, via affiliations, established in Europe (the European Chips and System Design Initiative (ECSI)) and Japan.

VSIA's aim was to reduce the confusion and design bottlenecks involved with IP reuse at the hard, firm and soft levels through the identification of *de facto* and *de jure* standards in use for IP development, exchange and integration, and the creation of new standards where nothing existed in the industry. Over time, there was also a realisation that many areas of IP use and reuse had not yet reached a point of standardisation, and what the industry needed was consensus on concepts, long before standards could be specified.   This led to the creation by VSIA DWGs of additional specifications and white papers, including system level model and functional verification taxonomies, several white papers on IP protection, definitions of concepts for system level interfaces, and a virtual component identification physical tagging standard. Fundamental work in delivering specifications and standards for on-chip bus attributes and virtual component interfaces, analogue/mixed-signal deliverables for AMS (analogue/mixed-signal) blocks, as well as signal integrity extensions, and virtual component transfer specifications, have been carried out by the DWGs.

More recently, in the 2001-2003 timeframe, VSIA has recognised the changing face of IP-based SoC design, and the growing importance of the concepts of platform-based design and the whole area of embedded software (SW) IP. In addition, the issues of IP quality and conformance have been growing. As a result, during this more recent period, VSIA has established new working groups in all of these areas.

Fundamental to the concept and hope of VSIA at the beginning in 1996, was the idea that IP could be created and packaged to this ideal of a standard set of interfaces and deliverables - the "virtual socket", whether delivered as soft - synthesisable RTL, firm - a netlist or placed netlist, or hard - GDS II (especially relevant for AMS IP components).   This interface or socket allows separation in time, space and commercial relationships between IP suppliers and IP consumers. At the same time, its existence lowers the risks of buying and integrating IP and would help foster the emergence of a commercial 3rd party IP industry by this lowering of technical risks and barriers. The method of integration assumed behind the VSIA concept was a block-based design and integration process - where IP blocks would be identified, procured and integrated on a block by block basis into an SoC.

IP creators would benefit; semiconductor and systems companies would benefit by lowering their design costs for application-specific standard products (ASSPs) and complex system designs and ASICs, giving them additional competitive sources for procuring new IP, especially in new growth markets. EDA companies would benefit through the identification of the standard formats which their tools would need to support in order to allow for efficient IP creation, exchange and integration. At the same time, new tool creation opportunities and markets would emerge.

2.2   The Reuse Methodology Manual (RMM) [8]

The RMM elaborated on the block-based design approach assumed in the VSIA concept, by developing and describing in detail a series of guidelines and rules for effective creation and reuse of individual soft (synthesisable RTL) IP blocks. It was based on the experiences gained by the two authors,

Michael Keating and Pierre Bricaud, and their colleagues, while participating in IP-based design projects at Synopsys and Mentor Graphics respectively. These guidelines and rules have been updated twice since original publication in 1998.

The RMM covers a number of useful topic areas, including, in its 3rd edition, outlines of system-on-chip design processes, rules and tools; macro block design process, including RTL coding guidelines, synthesis, and verification guidelines for soft macros; and hard macro design guidelines. It also includes concepts for packaging macros for reuse, system level integration and verification guidance, and data and project management suggestions. Guidelines vary from overall chip design suggestions (for example, chip-level clock distribution) to block guidelines (clock gating a block's operation) to individual design elements (clock gating at individual flip-flops).

2.3   The "Alba" Project

Since the beginnings of the SoC Revolution there have been several academic-industry-government collaborative projects which have attempted to establish or augment the capabilities of specific geographic regions to support, enhance and attract SoC design activities within their jurisdictions. These government-sponsored projects have taken many different forms, and have emphasised different concepts for IP-based design, SoC design, and System-Level Integration (SLI). The "Alba" project in Scotland [9] is one of these consortia, albeit the most ambitious and notable; there are many others. In Canada, for example, the Canadian Microelectronics Corporation [10] has emphasised the addition of a System-on-Chip research network [11] to its already well-established capabilities supporting university research and teaching in microelectronics and microsystems. In Taiwan, the "IP Gateway" sponsored by the Industrial Technology Research Institute (ITRI) [12] aims at strengthening IP-based design among the many small and large design houses in Taiwan, complementing the silicon foundry strengths of the country. In the U.S., the Pittsburgh Digital Greenhouse [13] in Pennsylvania and the project "Yamacraw" in Georgia [14] have taken a regional approach to fostering design activity and industrial development in their particular areas, while drawing on local university research activities at CMU and other Pittsburgh-area institutions, and Georgia Institute of Technology, respectively. The System integration and Intellectual Property Authoring Centre (SIPAC) in Korea [15] and SOCWare in Sweden [16] also have emphasised IP and IP-based design as a focus. We will discuss the Alba project in more detail.

One key focus in Scotland with project Alba has been the establishment of its three key pillars: that is, three sustaining institutions to support and foster SoC-based design and SLI. In Scotland, the Institute for System-Level Integration (ISLI) [17] was established as a unique consortium of the four major central belt universities with strong electronics engineering and computer science focus (Glasgow, Strathclyde, Edinburgh and Heriot-Watt) to develop and offer design tools and infrastructure for university teaching and research in SoC/SLI, as well as to provide related education, research and training. ISLI has made a concerted effort to update curricula, develop and deliver new courses, and offer industrial training in SoC-related topics. This included distance learning at the Scotland, U.K. and European levels.

A second pillar of the Alba project was the creation of a business infrastructure, the Virtual Component Exchange or "VCX" [18], to improve and promote IP exchange, and to develop solutions to the legal, business and economic issues associated with this area. The VCX is intended to be a structured market for IP exchange and trading under an accepted legal and business framework which would dramatically reduce the time required to deal with the legal, business and contractual issues for IP evaluation and acquisition for reuse.

The third pillar of project Alba was the creation of a dedicated Alba centre, funded by Scottish Enterprise (the government funding agency), to foster the development of a central SoC campus in Livingston, Scotland (about 15 miles west of Edinburgh). This would encourage the establishment of design centres within this campus who would make use of the ISLI and VCX infrastructure. ISLI is relied on to be a training vehicle for experienced staff needing upgrading in SoC design, and a source for trained new graduates. Via its trading floor, VCX is a source and supply of IP.

Successes for Alba in its early period of 1997-1999 included the establishment of a design centre by Cadence Design Systems, focusing on a variety of IC design projects including SoC and IP-based design. Motorola established an embedded software development centre, and other companies which also created design or development centres in the Alba campus included Epson, and Virtio.

The Alba project had an extremely ambitious aim - nothing less than a major economic transformation in Scotland's role in high technology business. It aimed to move Scotland from "screwdriver assembly" of electronic goods to playing an important and central role in the emerging area of SoC design, and establish a stable set of design firms for reasons other than government subsidies of a transient nature. Recognising the importance of synergies between its various elements, Alba put the role of education and research, front and centre in establishing the ISLI and in improving the Scottish university-level educational system in its ability to graduate highly-trained designers ready to work in the area of SoC.

In the next section, we will look at what has happened in the area of SoC in the last few years, including the Alba project.

3. THE REIGN OF TERROR: REALITY AND DISAPPOINTMENT 1999-2001

Revolutions progress through distinct phases, and the SoC revolution has followed this trend. Looking to the French Revolution as an example, we know that after the initial enthusiasm and social and political transformation, a "Reign of Terror" ensued in which the Revolution turned on itself, seeking traitors to the cause and engaging in an orgy of violence and death.

Thankfully, electronics has not been actually violent. But the area of SoC design, IP-based reuse and electronics design in general has gone through a "violent" upheaval, starting in 1999 and continuing through the electronics downturn of 2000-2002. This downturn, fuelled by the bursting of the "dot-com" bubble, the collapse of the grossly overvalued communications sector, and a fair degree of corporate and Wall Street shenanigans, has brought gloom, layoffs, and collapse to a large part of the industry. IC design has been no exception.

From the mid-1990's, various analysts from Gartner/Dataquest, iSuppli and IBS (Dr. Handel Jones) have produced a number of varying statistics on ASIC and ASSP design starts and design completions. Although the specific numbers vary widely, all these statistical series show a gentle decline in ASIC design starts or completions from the mid-1990's onwards, followed by a precipitous decline in the 2000-2002 period of between 30% and 60%. Forecasts for the future either show continued decline, albeit at a gentler rate, or stagnation.   If we look at the combination of ASICs and ASSPs (based on the argument of ASSPs substituting for ASICs in many designs) we see the same gentle decline, followed by precipitous decline, followed by stagnation or an anemic recovery in the 2003-2007 timeframe.

No matter whose statistics one looks at, the decline in IC design seems a marked phenomenon - and one only has to look at the dot-com and telecommunications crash and the enormous job losses, which have bitten deeply into design as well as manufacturing, sales and marketing, to realise that the industry has gone through a profound transformation in the last few years. And it is not at all clear that recovery will take us back to the point where we were.

When we think about this change, there are three possible sources for the decline in design starts or completions - an economic collapse; possibly, a transformation in design styles; and the enormous increase in Non-Recurring Engineering (NRE) charges for ASICs and ASSPs. More likely, it reflects a combination of all three factors. Electronics technology now allows the contents of several ASICs to be combined into one integrated SoC, or more highly integrated chipset or System-in-Package (SiP). So part of the decline in absolute numbers is no doubt due to a transition to more highly integrated SoC designs. The economic and performance imperatives at 180 and 130 nm demand such a transition.

However, the move to SoC design, which is real, has not been completely without tears. In the late 1990's a number of commentators began talking about SoC and IP reuse as a failure, based on both the failed early promises of the marketing "hype" around the concept, and a rush to embrace reuse without carefully preparing the infrastructure, methodologies, tools and flows to make it truly effective. In 2000, Jurgen Haase of SICAN (now sciWORX, now majority-owned by Infineon) gave a talk illustrating the public perception of the amount of IP reuse: growing exponentially from 1994 to 1997-98, followed by a crash due to over-"hyping" and subsequent disillusion, in 1998-2000 [19]. Interestingly, he laid the claim that the real business of IP reuse continued quietly behind the scenes to grow at a more measured pace from 1993 to 2001 and beyond, driven by the real successful design users who took a more orderly approach to incorporating IP reuse into their SoC development processes.

At IP 2001, a keynote speech by Jim Nicholas of ST Microelectronics asked the rhetorical question "System-on-Chip: more trouble than it's worth?" [20].   Of course, his answer was that it was worth the trouble - but only if approached using a certain design style we will discuss later. If we look at the IP market, we see that it has bifurcated over the years into two distinct sectors - the "Star IP" sector, primarily composed of processor cores and associated peripherals and buses - for example, ARM, ARC, Tensilica, PowerPC, OAK DSPs, and several others; and the "less-than-stellar" sector, consisting of relatively simple IP blocks, many of which are standards-based: for example, bus interface, packet encoding and decoding and simple peripheral control blocks. Of these two sectors, only the Star IP sector has been healthy - and indeed, recent failures and difficulties of a number of IP companies (BOPS, Chameleon, Morphics, Trimedia are all examples), indicate that during the downturn even stellar IP is no guarantee of market success. The less-than-stellar category has been subject to extremely rapid commoditisation and loss of commercial value due to low barriers to entry, no proprietary advantages for any one firm, and, as a result, has been a poor business for many entrants. The commercial, 3rd party IP market has never developed as the early advocates of the mid-1990's had hoped.

However, an internal IP market has definitely developed in many large semiconductor companies. A number of them (Motorola, ST Microelectronics, IBM, TI, Toshiba, Philips, Infineon and OKI are just some examples) have made considerable internal investments in IP-based design infrastructure and encouraged significant IP reuse between groups within the company. Many of these internal efforts utilised reuse standards from the VSIA as one basis for their internal standards and systems for IP. They also added substantial proprietary and company-specific concepts to the IP-based design systems.

It has been within large companies, rather than via a large and growing 3rd party IP industry, that IP-based design has achieved the most success. However, we must remember that some Star IP companies - for example ARM, have been conspicuously successful and may be an example for many others to follow.

If we look to the Scottish Alba project, we can see that after some initial success and the establishment of some design centres along with the ISLI and VCX, the picture in 2002-2003 is distinctly mixed. The electronics slowdown has not been conducive to long-term investments in new design centres, although it has encouraged a rapid growth in student applications for advanced training in SoC/SLI at the ISLI. Similarly, the VCX has managed to survive, but has never taken off as the primary world centre for IP trading. Alba has not failed; but it, and the promise of SoC design, has not delivered on its initial, over-inflated, early expectations either.

Summing up the state of IP-based design in mid 2002, Richard Goering of EE Times made a very interesting set of observations, which point to a very different approach to IP-based design that we will discuss more in the next section [21]:

"… commodity IP accounted for just $114 million in 2001, a fraction of the $892 silicon IP market. … Synopsys and Mentor will now dominate that market. … Buying commodity IP will be more akin to buying a T-shirt at Kmart than walking into a small, locally owned T-shirt shop. … The lion's share of the IP market is what is called 'star' IP. … You might think of buying star IP as somewhat like buying a colorful, expensive tie from the Tie Rack, an international chain specializing in ties and scarves. … But the latest trend among IP suppliers is to provide more than just ties-I mean, cores-and to offer 'platforms' that also include peripherals and a predefined bus architecture. … I think of buying platform IP as similar to walking into the Men's Wearhouse, a heavily advertised West Coast chain where a salesperson will help you pull together a coordinated 'look' with suit, shirts and ties that's customized for you."

4.   THERMIDOR: THE PLATFORM TRANSFORMATION, 2000-2002

After their Reign of Terror, revolutions usually go through a transformation and period of quietude, during which the foundations of the long-term subsequent next state are put into place. In the French Revolution, this period was called "Thermidor" - after the arrest and execution of Robespierre, who was primarily responsible for the Reign of Terror period.

In IC design, over the last several years, we have seen a shift in the attitude and approach to IP-based design from a large number of companies. That shift has included a move to "Platform-based Design" of SoC. What is Platform-based Design? The definitions are legion.   Several of us put one stake in the ground in our 1999 book *Surviving the SOC Revolution* [22]. More recently there has been the development of more formal definitions, recognising platforms at multiple levels of design and abstraction [23]. The VSIA Platform-Based Design Development Working Group (PBD DWG) has come up with two relevant definitions in its PBD Taxonomy, still under development as of the spring of 2003 [24].

According to the VSIA working group, a ***platform*** is "An integrated and managed set of *common features*, upon which a set of products or product family can be built. A platform is a virtual component (VC)."   They then define ***platform-based design*** as "An *integration oriented* design approach emphasizing systematic reuse, for developing complex products based upon platforms and compatible hardware and software VCs, intended to reduce development risks, costs, and time to market."

These general definitions are pretty reasonable, and can be applied at levels beyond the SoC.   Based on many of these concepts, we can propose a more precise definition in this book as:

"***Platform Based Design*** is an organised method to reduce the time required and risk involved in designing and verifying a complex SoC, by heavy reuse of combinations of hardware and software IP. Rather than looking at IP reuse in a block by block manner, platform-based design aggregates groups of components into a reusable platform architecture."

One must note that platform-based design is *not* the only approach used in SoC design.   It is also not the only approach to heavy IP-reuse based design of SoCs. As discussed in [22], there are several approaches, including more *ad hoc* block based design and integration methods. The choice of the SoC design approach is based on the application requirements - the exact tradeoff of cost, performance, power consumption, design risk, and time-to-market. But it is a very reasonable approach for many SoC designs, demonstrated by the significant gains in popularity over the last several years.

When we pulled this book together, we contacted several potential authors from several companies leading the world in SoC design. We had no pre-conceptions as to the particular SoC design styles used and advocated by these contributors. As the chapters came back, it was both interesting and gratifying to see how many of these companies have made platform-based design and its related concepts a centrepiece of their SoC design strategy. As you read more of this book, readers will note a considerable influence of the platform approach in many (but not all) of these chapters discussing SoC success. Other design and verification approaches for SoC are discussed, for example, in [25, 26].

The industry has taken to platform-based design for many SoC projects for several reasons. The most important among these are:

- It is the next logical step in IP reuse, moving up from *ad hoc* block reuse to the reuse of aggregates of IP blocks and an integration architecture. As such, it greatly reduces design effort and risk, thus improving time to market.
- Rapid design derivatives become possible, allowing optimization and tailoring of products to very specific application needs and markets.
- Platforms are a way of capturing and reusing the best architectures and design approaches found for particular types of products and markets. In general, there are only a few optimal architectures for particular application domains, and platforms crystalise and harden these approaches for reuse by others. They thus also serve as a transmission mechanism from more experienced design teams and architects to less experienced designers.

Platform-based design is a strong reality in SoC today, and many believe it is the most powerful IP-reuse based approach to SoC, now and in the future. Indeed, platforms can be considered in their more abstract sense as "a co-ordinated family of hardware-software architectures, satisfying a set of architectural constraints that are imposed to allow the reuse of hardware and software components" [23]. This can be called a System Platform, which is a reconciliation point allowing multiple applications from a product design domain (the application space) to be mapped into a particular platform instance drawn from the architectural space; or a single application to be mapped, potentially, to multiple different platform instances in order to find the optimal target for the design. This latter exploration of mappings is what is known as design space exploration.

We can approach platforms in several different ways. The hardware-centric view considers a platform as consisting of a fixed HW-SW kernel, along with a variable portion that allows the platform to be tailored to specific application domains (Figure 1-2). The kernel consists of a number of scalable physical architectures for clocking, power, interface, on-chip bus, test and timing.   The variable region consists of variable SW and HW (standard cell logic, reconfigurable logic, and options in between, such as metal-programmable gate array).   The software view of a platform essentially is a programmer's model: an application programming interface (API) that abstracts away the underlying hardware detail of the platform and presents to the system and software designer just enough of that detail to allow them to map their applications to the platform, while taking advantage of the hardware. In other words, the API sits on top of the device drivers, BIOS, and much of the real-time operating system (RTOS) as well as the processor (viewed via a compiler, debugger, etc.) and network stacks. See [22, 23] for much more on this.

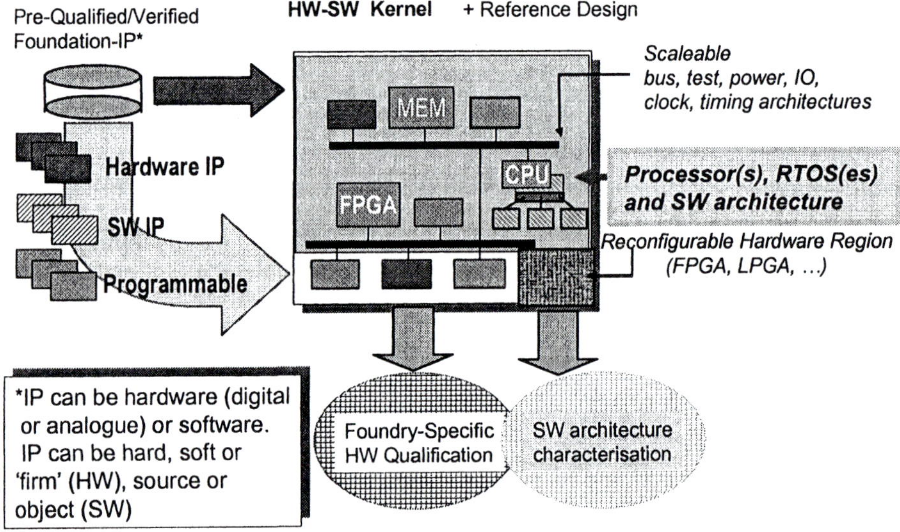

Figure 1-2. A hardware-centric view of a platform

The final view of a platform could be considered to be a "fabric-centric" view, which concentrates on the platform as a delivery vehicle for several different forms, or "fabrics" for design - analog HW, digital standard-cell HW, custom HW, reconfigurable logic, gate arrays, hard IP cores, and SW expressed in several forms (object, source, APIs, service layers). We can then think of a platform as an architecture built on a mosaic of different fabrics, which provide several varying layers of design abstraction.

There are of course several different platform styles in use in the marketplace, and the subsequent chapters in this book discuss many of them. This includes full application platforms, reconfigurable platforms, and processor-centric platforms [27]. Full application platforms, such as Philips Nexperia (Chapter 4) and TI OMAP (Chapter 5) provide a complete implementation vehicle for specific product domains. They deliver to design groups libraries of hardware and software components, along with several mapping examples via reference designs. However, these platforms require the provider to make the greatest modeling effort and take the most investment risk. Processor-centric platforms, such as ARM PrimeXsys (Chapter 9) concentrate on the processor, its required bus architecture and basic sets of peripherals. They also deliver basic software drivers and routines, along with RTOS ports and a software development environment. Being more generic than the full application platforms, they take less effort to develop and have lower investment risk, but more work is required to turn them into a derivative product.

Reconfigurable platforms such as the Xilinx Platform FPGA (Chapter 7) and Altera's SOPC (Chapter 8) deliver hardcore processors plus reconfigurable logic along with associated IP libraries and design tool flows. They represent different tradeoff points along the cost-performance-power axis than the standard-cell based platforms.

We also distinguish platform users [27]. First, the power users, who differentiate their end products based on a platform at all levels including HW and SW, by modifying and developing custom HW and SW components. Second, the platform differentiator works at the application level, by developing SW and choosing IP from existing libraries. Finally, the complete package user, who wants a complete HW and SW packaged solution, and differentiates primarily through modifying application software.

Finally, we should recognize that the platform concept recurs at multiple levels of the product development "stack". The system platform mediates between system product developers and SoC architects. Similarly, the silicon implementation platform mediates between the architecture instance for a particular design and the silicon implementation fabric offered by a particular technology and associated design kit [23].

5.   L'AVENIR: THE FUTURE IS STILL RADIANT AFTER ALL?

We now briefly consider the future of SoC design. After the French Revolution, the future of France became bound with a new Emperor: Napoleon. We must speculate on which future Napoleon of SoC will emerge for our design community in the next few years. This is extremely speculative. The last chapter in the book discusses the impact of new process technologies on SoC design in more detail. Here we will consider the possibilities opened up by new SoC design architectures.

We have seen the emergence in recent years of a new kind of design architecture and corresponding "SW compilation" model, where algorithms captured in classical SW form as sequential C are automatically compiled into a combination of SW running on more conventional processors, and specialised HW adjuncts which speed up critical parts of the processing. This is enabled by the control interfaces between HW and SW, which are also generated. The processors range from fixed RISC to complex custom-compiled VLIW. The hardware ranges from ASIC-style standard cell implementations to the use of reconfigurable logic. Sometimes an intermediate form of "HW adjunct" is involved, in configurable embedded processor cores, where special application-specific instructions can be created via microcode which provide significant speedup for critical parts of SW.

Companies and research groups involved in these kinds of capabilities include HP Labs with its "Program-In, Chip-Out" (PICO) project [28], UC Berkeley with GARP [29], ARC Cores and Tensilica with instruction configuration of embedded processor cores, and small startups such as Proceler. The new highly programmable platforms involving embedded processors and large amounts of reconfigurable logic, such as Xilinx Virtex II Platform FPGA and Altera System-On-a-Programmable Chip (SOPC), would seem to be natural targets for these kinds of capabilities because of their extreme flexibility, but such approaches would also work well with more conventional custom and semicustom approaches.

One interesting point to note is that essentially such approaches attempt to rediscover the natural concurrency within an algorithm – concurrency that can be accelerated by HW implementation using pipelining and parallel HW. It is interesting to speculate if these techniques, combined with a more natural form of expression not found in sequential C programmes, might give superior results. For example, dataflow diagrams of algorithms, in which networks of atomic functional units represent the algorithm and the parallelism is more naturally exposed, may be a better notation. Some companies are emerging to commercialise academic research in the direct compilation of algorithms to HW, such as Accelchip for DSP algorithm implementation from The Mathworks' Matlab and Simulink to FPGAs [30]. The Berkeley Wireless Research Centre also has an interesting approach with its direct mapping concept from Simulink to a HW realisation [31, and Chapter 10].

5.1   Networks of Flexible Processors: The Mapping Problem

Future SoC architectures are increasingly being suggested to consist of networks of flexible and configurable processing units, linked by a configurable on-chip communications fabric, or network [32].

The general problem of mapping a number of possibly-concurrent algorithms and functions onto this sea of processors, given the opportunity to configure each processor, to configure HW adjuncts to processors, and to configure the communications network, all dynamically, in the face of a variable and very dynamic processing load, is an extremely challenging research and industrial problem.   This is critical to the future of SoC. Indeed, this is arguably the *real* system-level design problem for SoCs, and the one that will occupy both researchers and commercial designers and the EDA industry for some considerable time to come.   As always, the complexity of SoCs is outstripping the arrival of new tools.

5.2   Does System-Level Design of SoC = SW?

The combination of mapping function to architecture both in the narrow sense of a single algorithm mapped to a processor and adjunct HW, and the broader sense of many algorithms mapped to a flexible sea of processing elements, presents system level design with a profound dilemma. Increasingly, those who use these complex platforms and SoCs will be SW developers, and systems designers with a SW orientation.   They will want the detailed HW aspects of the target SoCs and platforms to be hidden from them. If we follow this line of reasoning to its fullest extent, then system-level design increasingly will be seen as SW design. Functionality of systems will be described in a software form.   Highly programmable platforms will offer a variety of system resources on which this functionality will be mapped – but whether a function is mapped to HW (fixed or reconfigurable), or SW, or a combination, will be built into SW-driven design flows using compilation paradigms.   Design space exploration may be equivalent to the "-O" (for optimisation) flags on compilers, presenting designers with a variety of implementation alternatives offering Pareto-optimal combinations of performance, power and cost. The optimisation criteria may come from a variety of (hidden) HW-SW architectural and performance models, but designers will not need to directly manipulate them, nor understand the details of HW implementation; the tools will do it all. The concept of "SW Washing Machines" from Francky Catthoor and others at IMEC [33] fit into this approach, in that they will optimise the software implementations of system applications, especially legacy SW, in a target platform-independent way. These can then be optimally mapped onto a particular target platform – a network of flexible computational resources. In this scenario, it will not be easy to distinguish a system-level design flow from today's SW development flows.

It is thus very likely that although considerable work is required to research and deliver tools which manage and optimise the complex mapping problems involved in future SoC systems, to a large extent this will be hidden in optimising compilers. As a result, these new capabilities will be unsung and unremarked by designers; extracting commercial value from them will be a profound challenge to the industry.

6.   CONCLUSIONS

We have seen in this history of SoC design that it has gone through several distinct stages, from excitement through euphoria through disillusionment, panic, transformation, and emerging from research labs and leading architects, some very interesting possibilities for future SoC architectures.

But the most important lessons from this history are: SoC design is a reality today; there are many leading groups successfully using this approach, and if you read on in this book, you will learn many of their ideas, approaches and concepts.

Please both *enjoy* and *learn* from the insights which follow.

REFERENCES

1.   Gordon E. Moore, "No Exponential is Forever: But 'Forever' Can be Delayed!", *Proceedings of ISSCC 2003*, pp. 20-23.
2.   G.E. Moore, "Cramming more Components onto Integrated Circuits", *Electronics*, Volume 38, number 8, April 19, 1965.
3.   International Technology Roadmap for Semiconductors (ITRS), 2001 edition.
4.   International Technology Roadmap for Semiconductors (ITRS), 2001 edition, "Design" Chapter - Figure 15, "Impact of Design Technology on System Implementation Cost", and Appendix: "DT Cost and Value".
5.   Andrew Kahng and Gary Smith, "A New Design Cost Model for the 2001 ITRS", *International Symposium on Quality Electronic Design 2002*, pp. 190-193.
6.   R. Goering, "Design reuse will cut costs in half, study says", *EE Times*, September 28, 1998. Refers to study on IP return on investment by Michael Keating of Synopsys. R. Goering article on the web at URL:
      http://www.eetimes.com/news/98/1029news/design.html
7.   Virtual Socket Interface Alliance, on the web at URL: http://www.vsia.org. This includes access to its various public documents, including the original Reuse

Architecture document of 1997, as well as more recent documents supporting IP reuse released to the public domain.

8. Michael Keating and Pierre Bricaud, *Reuse Methodology Manual for System-on-a-Chip Designs*, Kluwer, 1998 (1st. Edition), 1999 (2nd. Edition), 2002 (3rd. Edition).

9. http://www.albacentre.co.uk/

10. http://www.cmc.ca/

11. http://www.cmc.ca/news/bulletins/socrn_newsletter/socrn_newsletter_1.2.pdf

12. http://www.taiwanipgateway.org/index.jsp

13. http://www.digitalgreenhouse.com/

14. http://www.yamacraw.org/

15. http://www.sipac.org/

16. http://www.socware.com/default.htm

17. http://www.sli-institute.ac.uk/

18. http://www.thevcx.com/

19. Jurgen Haase, "Virtual Components - from Research to Business", *Forum on Design Languages 2000*, Virtual Components Design and Reuse forum keynote address.

20. Jim Nicholas, "System-on-Chip: more trouble than it's worth?", *IP 2001* keynote..

21. Richard Goering, "Shopping for Silicon IP", *EE Times*, August 6, 2002.

22. Henry Chang, Larry Cooke, Merrill Hunt, Grant Martin, Andrew McNelly, and Lee Todd, *Surviving the SOC Revolution: A Guide to Platform-Based Design*, Kluwer Academic Publishers, 1999.

23. Alberto Sangiovanni-Vincentelli and Grant Martin, "Platform-Based Design and Software Design Methodology for Embedded Systems", *IEEE Design and Test of Computers*, Volume 18, Number 6, November-December, 2001, pp. 23-33.

24. VSIA PBD DWG, *Platform-based design definitions and taxonomy*, working document, Revision 0.90, March 13, 2003.

25. Rochit Rajsuman, *System-on-a-Chip Design and Test*, Artech House, 2000.

26. Prakash Rashinkar, Peter Paterson and Lenna Singh, *System-on-a-Chip Verification : Methodology and Techniques*, Kluwer Academic Publishers, 2001.

27. Grant Martin and Frank Schirrmeister, "A Design Chain for Embedded Systems", *IEEE Computer*, Embedded Systems Column, March 2002, pp. 100-103.

28. Vinod Kithail, Shail Aditya, Robert Schreiber, B. Ramakrishna Rau, Darren C. Cronquist and Mukund Sivaraman, "PICO: Automatically Designing Custom Computers", *IEEE Computer*, September 2002, Volume 35, Number 9, pp. 39-47.

29. T.J. Callahan, J.R. Hauser, and J. Wawrzynek, "The Garp architecture and C compiler", *IEEE Computer*, April 2000, Volume 33 Issue 4 , April 2000 pp. 62-69.

30. M. Haldar, A. Nayak,, A. Choudhary, and P. Banerjee, "A system for synthesizing optimized FPGA hardware from Matlab", *Proceedings of ICCAD 2001*, pp. 314 -319.

31. Davis, W.R.; Zhang, N.; Camera, K.; Markovic, D.; Smilkstein, T.; Ammer, M.J.; Yeo, E.; Augsburger, S.; Nikolic, B.; Brodersen, R.W. "A design environment for high-throughput low-power dedicated signal processing systems", *IEEE Journal of Solid-State Circuits*, Volume 37, Issue 3, March 2002, pp. 420-431.

32. Hugo De Man, "On Nanoscale Integration and Gigascale Complexity in the post .COM world", Keynote address, *DATE 2002*, Paris, March, 2002, slides available at URL: http://www.date-conference.com/conference/keynotes/deman/deman_slides.pdf/.

33. Chris Edwards, "Washing machines the key to low-power processing", *EETimes*, 7 March 2002. URL: http://www.electronicstimes.com/story/OEG20020306S0001/.

Chapter 2

SOC DESIGN METHODOLOGIES

Henry Chang
Architect, Cadence Design Systems, Inc.

Abstract: SoC design incorporates the complete panoply of complex IC and embedded software design issues, including their relationships to other design tasks such as chip packaging and printed circuit board design. How one sets up one's design methodology becomes one of the most critical factors for success. This chapter presents an overview of the breadth and scope of the design steps required for the design of an SoC. It begins with a discussion of platform based design as one of the overarching approaches to "winning" with SoCs. It then describes the basic steps- system design, embedded software design, functional verification, hardware IC design, and analog/mixed-signal in SoCs. This is followed by a discussion of some of the often forgotten other components- the infrastructure required, the interfaces between the design teams, and the "meta methods" or design management related tasks. This chapter also sets the stage for the remainder of the book often referring to later chapters for "real" design experiences and more detailed discussions.

Key words: Design methodology, platform based design, system design, embedded software, verification, IC design, analog/mixed-signal, infrastructure, interfaces, meta-methods.

1.   INTRODUCTION

The design of an SoC is a formidable challenge requiring a sizeable design team and many design disciplines.   This chapter provides an overview of the many design tasks and disciplines required for successful SoC and SoC platform based design. It will highlight the key steps and key issues associated with those steps to give the reader a one chapter overview of an SoC design methodology. This chapter also serves as a backdrop to the remaining material found in this edited volume. Where there is elaboration on a topic a chapter reference will be provided. Also, where there is more detail on that topic in the original *Surviving the SoC Revolution* book [1], a chapter reference to that text will also be provided.

Figure 2-1 gives a simplified view of an SoC design methodology, showing the basic design tasks required for an SoC design. The subsequent sections will describe each of the tasks. Section 2 will describe how to build a platform. Section 3 explores system design. Section 4 will describe the issues in embedded software development. Functional verification is covered in Section 5. The hardware design of the IC and design for test will be discussed in Section 6. Section 7 describes the aspects of Analog/Mixed-Signal (AMS) IP block integration onto SoCs. Section 8 discusses some of the infrastructure elements required focusing on the IP database.

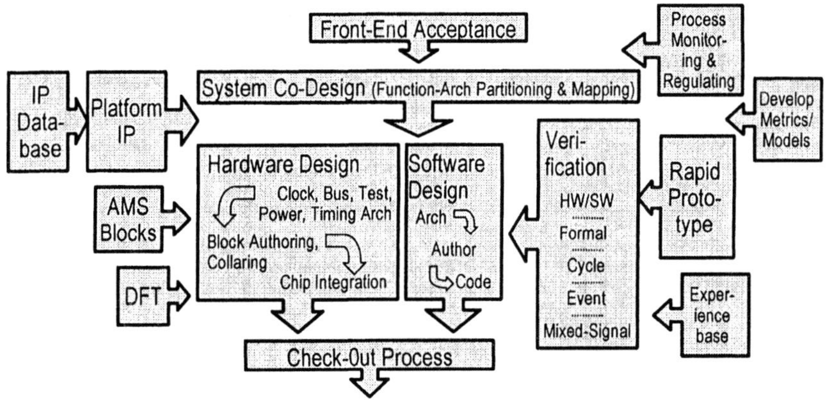

Figure 2-1. Overview of Steps in an SoC Design

Besides just the tasks, Section 9 describes the interfaces and cross-disciplinary challenges between the tasks. Finally, Section 10 describes the "meta-methods" or design management related tasks in the figure, including front-end acceptance, the check-out process, process monitoring and regulating, the development of metrics and models, and the use of an experience base.

2.   PLATFORM BASED DESIGN METHODOLOGY

Figure 2-1 presents a "flat" set of design tasks for SoC Design, meaning that for every SoC design that is to be done, one has to repeat all of the steps from beginning to end. If each SoC design is unique, then this is probably fine. However, for designs that are similar, this may be quite inefficient. We view platform based design as an overlay methodology on this base SoC approach, where "design step reuse" is utilized in the design process.   In other words, all of the same steps are required, but by performing many of the steps only once across many similar SoCs, overall design efficiency is improved.

There are four basic steps to using a platform based design methodology. This is shown in Figure 2-2. The first step is to define the methodology for designing a platform, i.e. planning how to design the platform. The next step is to actually create the platform using the methodology defined. The third is then to define how to use that platform, in order to create a derivative product. Finally, the fourth is to actually create a product or a family of products.   Thus, the advantage is obtained because Steps 1-3 are only performed once for an entire product family. Only Step 4 is repeated for each SoC design.

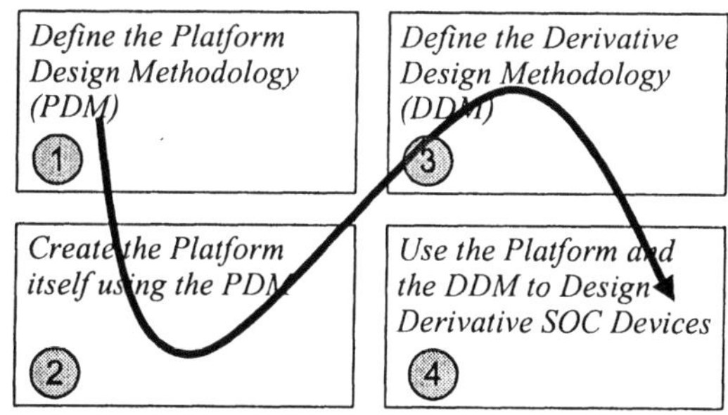

Figure 2-2. The Four Basic Steps to Platform Based Design

Chapter 6, Section 4.1 (from IBM Microelectronics) also describes two primary methodologies- "a methodology for Platform Creation" (Step 1); and "a Methodology for Platform Based Design" (Step 3).   Chapter 9, Section 2.2.1 (from ARM) expands the 4 step methodology by adding additional steps to Step 2.   They have "IP Licensees" who can further enhance the first Step 2 output beyond what the original "IP Creator" provides. Chapter 7 (Xilinx) describes a scenario where the semiconductor vendor provides the hardware fabric (in this case the FGPA) on which one of their customers then builds an application specific platform to complete Step 2. This is then delivered to the end customer who completes Step 4.

Figure 2-3 provides more detail on the Platform Design Methodology. The focus of this methodology is to define what is going to be reused in the derivative designs. It stops short of final implementation. Depending on how much is to be reused, different aspects could be driven closer to implementation. For example, if we know that all of the products in the product family will have to meet certain I/O standards, and we know that many products will be fabricated on the same manufacturing process, then the I/O hardware blocks may be taken all the way to layout. However, if there are aspects which will appear in some versions and not in others, one may stop at the Register-Transfer Level (RTL) for some blocks and not go to layout. It should be noted that though the platform does stop short of implementation, it is typical that as the platform is being developed it *is* taken completely to implementation and "taped out" to help verify the platform, serve as a reference design, or even be the first product.

In the platform design methodology, the bulk of the time is spent at the system definition level making certain that the platform can be used in the many applications required. The primary inputs to platform design are derived from the product family requirements as well as requirements dictated by standards. More and more, standards make up a large bulk of the requirements when defining a platform.

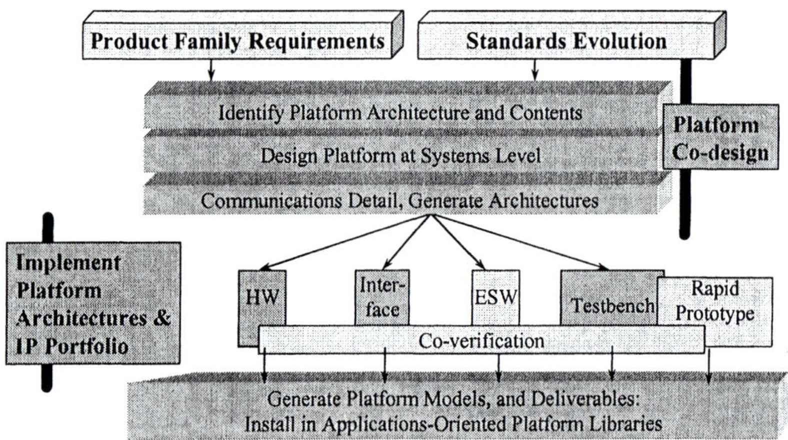

Figure 2-3. Platform Design Methodology

Following the inputs, a whole set of platform co-design steps are employed where the platform architecture and contents are identified. The platform is designed at a system level, identifying the platform architecture and contents. This is followed by the creation of the communication details and the generation of the detailed architecture. These steps will be described in more detail in the subsequent sections. Following this there is the design of the hardware (HW), the embedded software (ESW), and the interfaces, to the extent that these can be reused in most of the derivatives. Rapid prototyping environments can also be developed as well as the creation of a co-verification strategy across these domains. The result of this process is a platform. Chapter 4, Section 2 (Philips) and Chapter 5, Section 2 (TI) both provide unique perspectives on the different levels of abstraction for platforms. Chapter 9, Section 8 (ARM) provides an excellent example of what needs to be delivered with a platform.

Now we describe the derivative design methodology or how one uses the platform. This is shown in Figure 2-4. Specific product requirements now need to be considered and compared against the platform libraries available. We define a "front-end acceptance" process to evaluate the inputs for quality and completeness (see Section 10). We modify the platform as needed in system design, and we refine the links to implementation.

Unlike the design of the platform we now must actually finish the entire design, completing the HW, the ESW, and the interfaces. The result goes into the post-design phases of fabrication, ESW assembly, burning the embedded software in ROM (if necessary), integration of all of these components, and debug.

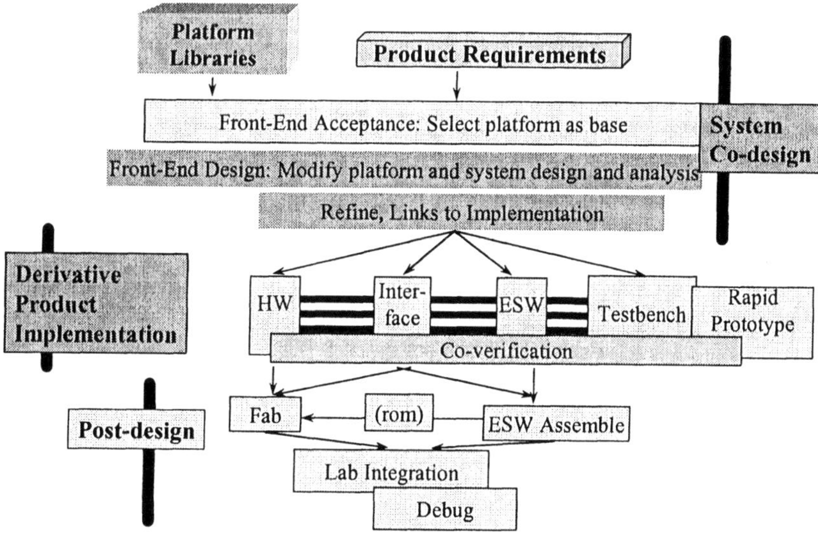

Figure 2-4. Derivative Design Methodology

Chapter 6, Section 4.2 (IBM) presents a case study on how the use of platforms greatly reduced their time for derivative designs.

3. SYSTEM DESIGN

The emphasis in system level design is the definition of system functionality, which models the product application, the definition and modeling of an appropriate architecture on which the functional model will be mapped, and mapping and analysis of the suitability of the architecture as a vehicle for realizing the application. System-level design is fundamental to effective SoC design, and is particularly effective when married with platform-based design concepts [2].

The essential concept is to allow exploration of design derivatives at the system level of abstraction, rather than the more traditional RTL and C level. Function-architecture co-design is a modern form of design space exploration that goes beyond hardware-software co-design [3, 4], to permit a wider range of design tradeoffs. When platforms are modeled at the system level with appropriate support for design space exploration, the creation of a derivative design and its validation can be done rapidly, with a high probability that the derivative design implementation will be successful [5].

This approach is shown in Figure 2-5. Step 1 is to describe the system function. Step 2 is to explore a set of various system architectures for the function. Step 3 is to do a mapping between the system function and architectures. Performance simulations can be done to see if particular architectures are appropriate for the system function in question. Chapter 4, Section 5 (Philips) describes some of the challenges to system architecture design and ESW scheduling. Chapter 8, Section 4 (Altera) also looks at system level performance requirements in the context of reconfiguring the FPGA for better overall system performance. The communications are refined, and Step 4 is the flow to implementation.

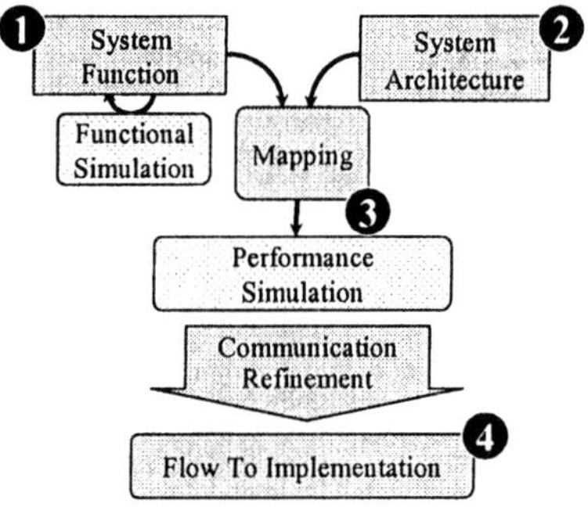

Figure 2-5. System Design Flow

Different approaches are possible, of course. Chapter 9, Section 2.2.2 (ARM) shows a variant of this. Chapter 10, Section 3 (BWRC) also provides a system-level design flow. One point is clear, however, that as design architectures become more complex (e.g. network of processors on a chip with complex memory and busing structures), this process becomes more "artistic" utilizing much creativity from the part of the design architect, and thus the process becomes highly iterative.

System level design also incorporates capabilities for specifying, modeling, simulating, and analyzing dataflow algorithms to solve signal processing problems in communications and image processing. This includes capabilities to model the complete system environment such as channel noise, reflective environments, and generalized control. Chapter 4, Section 5.5 (Philips) also describes some of the differences between datapath and control path operations in the form of software code differences.

As part of the flow to implementation another set of decisions that must be made at system design is around the area of SoC implementation fabrics. New options are available such as the highly programmable platforms based on a combination of fixed processor cores and reconfigurable logic. This is described in great detail in Chapters 7 (Xilinx) and 8 (Altera). As ASICs add reconfigurable logic regions for maximum flexibility (also described in Chapter 7, Section 3.4.1); as FPGAs become "platform FPGAs" with fixed processor and memory cores; and as new reconfigurable logic architectures emerge (e.g. metal programmable gate arrays), the boundary between the ASIC and FPGA design styles begins to blur. Ultimately, a hybrid of the two seems destined to become the optimal implementation fabric for a host of SoC applications. Perhaps a more interesting question is where the dividing line will be drawn between the hybrid, highly programmable platforms, and the more cost-optimized, less flexible, custom design approaches to SoC. Clearly some applications may never lend themselves to this approach but high economic rewards will accrue to those who choose the right path in each domain.

More discussion on system design, the design of communication networks, and implementation fabric trade-offs can be found in Chapter 4, 5, and 6 respectively in *Surviving the SoC Revolution* [1].

# 4.		EMBEDDED SOFTWARE DESIGN

Embedded software (ESW) is a growing portion of all embedded systems and the SoCs which go into them. Anecdotal evidence suggests that ESW development is now two to three times the effort compared to hardware design, and that the effort continues to grow faster than the hardware effort. A good overview of the issues involved in the embedded software area for SoC can be found in [6]. Key issues include:

– Ensuring that the right tradeoffs are made as to which part of the
   application processing is to be done in ESW running on a processor,
   which type of processor (microcontroller or digital signal processor
   (DSP)), which in dedicated digital HW, dedicated analog HW, and RF;

- Verifying the implementation of the ESW within a complex system environment consisting of these several domains;
- Ensuring the portability of ESW across future product evolutions, choices of real-time operating systems, processors, middleware, application stacks, etc. The legacy issues of ESW are always daunting.

Embedded SW clearly is different than application software on a personal computer (PC). PC software is not real-time and offers no guarantee of responsiveness. Reliability and ease of use have not been issues, especially considering the history of the prevalent operating systems and applications running on it. PC software has bloated considerably over time, relying on the underlying HW platforms (the PC platform moving from x86s to Pentium I, II, III, IV and beyond) to offer the vastly increased disk space, memory size and processing power in order to keep pace with software feature bloat. Since overall software size and runtimes are not particularly critical, applications developers have been able to use the latest tools and abstractions (object-oriented programming, Visual C++, for example) in order to achieve productivity. Finally, since the underlying platform is relatively fixed or moves in slow increments with reasonable amounts of backwards compatibility, porting of PC software is not a huge issue, although still a source of pain.

Contrast this with embedded software in real-time systems. Here, real-time characteristics are critical. Constraints range in importance from weak to very strong, and deadlines are often very hard, especially for safety-critical systems such as automotive that demand guaranteed responsiveness. Consumer demands for reliability and ease of use are much more important to the success of an embedded product. Cost considerations imply very tight constraints on software size, latency and responsiveness, with extreme attention paid to memory size and footprint, processor class, and assembly coding is still very much used especially for DSPs to guarantee performance. Finally, porting of highly-tuned SW from one platform to another is very much an issue.

Figure 2-6 illustrates one possible architecture or layering for embedded software in a platform-based design approach [7]. The most important concept shown here is not the specific layering, but the concept of a "Platform API" which hides the details of the underlying HW architecture and the low-level ESW from the applications designer, without limiting their ability to optimize the resulting code. The whole area of effective mechanisms for ESW development in a platform-based SoC context is under active study [8, 9]. Chapter 4, Section 4.1 (Philips) describes a similar software layering approach.

To add a twist to this, Chapter 7, Section 5 (Xilinx) describes a methodology for reconfiguring the hardware (on a FPGA) while the software is running to greatly add to the flexibility of an SoC Platform. Chapter 8 (Altera) also describes a scenario where if the ESW is found to run too slowly, then dedicated hardware can be added to accelerate key operations.

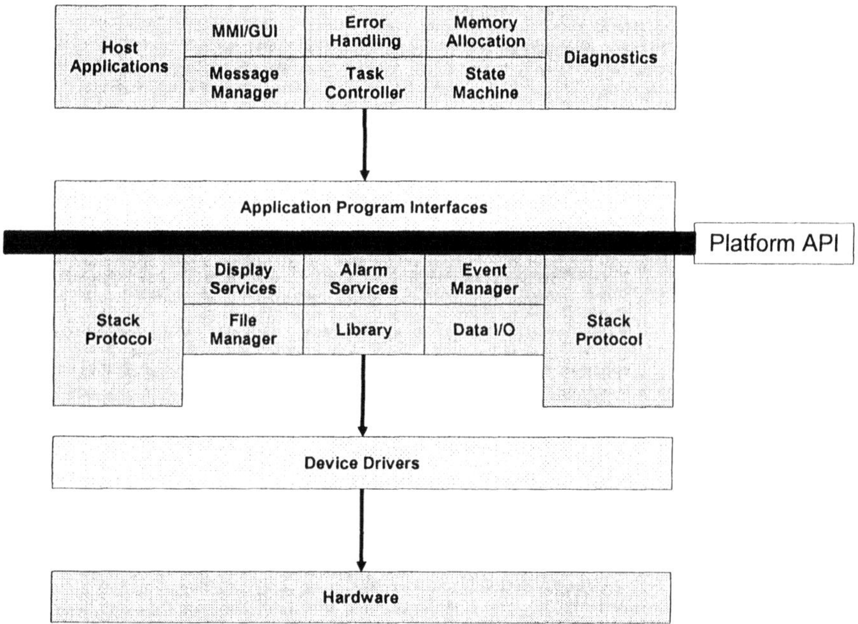

Figure 2-6. Embedded Software Architecture Example

In general, effective solutions to these issues revolve around careful architecting of ESW to layer it into a small amount of highly-optimized platform-dependent code, and a much larger amount of well-structured platform-independent code which can be more easily ported to new processors, real-time operation systems (RTOSs) and applications. This approach has been successful on a number of complex applications in the communications and automotive fields. Details on how to write embedded software for SoCs are described in Chapter 9 of *Surviving the SoC Revolution* [1].

5.   FUNCTIONAL VERIFICATION

It has been repeatedly stated in recent years that complex IC (hardware) designs split into approximately 30%-40% design effort and 60%-70%

verification effort. SoC design with its additional system-level complexity only compounds this issue. However, a carefully chosen hierarchical and multi-level verification strategy involving modern tools can go a long way to ameliorate this problem. Chapter 4, Section 3 (Philips), and Chapter 9, Section 3.4.2 (ARM) have very good descriptions of the levels of abstraction they use to model RTL and above. A list of abstraction levels is also defined in Chapter 7 of *Surviving the SoC Revolution* [1]. Verification is so critical that various approaches are described in the remaining chapters. Chapters 6 (IBM Microelectronics) and 10 (BWRC) focus on the topic of verification. Chapter 9, Sections 5-7 (ARM) describe all that is delivered to support their customers' verification needs. The maze of current and new verification approaches, going beyond simple simulation, can be overwhelming for users to understand. The chapters cited provide good insight into the approaches being taken. Also, two good guides to SoC functional verification are [10] and [11].

We cannot begin to describe all of the approaches here. We will restrict our discussion of verification to just the block-based hardware level, because it has many of the features of a hierarchical verification approach that we are advocating. Here it will suffice to say that the emphasis in verification for SoC is a divide and conquer strategy that exploits the underlying hierarchical nature of the SoC design; and a transaction-based verification strategy that moves testbench stimulus, response and checking up from the Boolean signal level to more complex data types and system level transactions.

When verification strategies map onto the underlying SoC hierarchy, it is possible to verify designs block by block using a unit test concept, and then using selectively either the full block model, or a bus-functional model equivalent, within the overall SoC verification model. Testbenches similarly can be built on a unit, block basis, and reapplied within the overall SoC model. This is made much easier if the testbenches are architected to use the notion of high level transactions that are "natural" for the SoC. For example, processor and peripheral blocks communicate with on-chip buses via various bus read/write and acquire/release transactions. It is thus easier to create testbenches that describe tests at this level and "protocol-convert" them to bit-wise signals, than to write them all at the bit-wise level. It is also possible to write higher level models for the blocks to allow simulation at this level. This method greatly facilitates testbench reuse.

One implementation of this approach is to use verification wrappers. This is illustrated in Figure 2-7. The terms "transactors" and transaction verification models (TVMs) are also often used to describe the wrapper concept [12, 13]. These wrappers are put around the major blocks. The "platform foundation block" represents the hardware part of the platform. The other blocks represent the new intellection property (IP) blocks that will be added to the derivative design. These wrappers can be used to apply and capture test suites. They can drive these test suites into their associated blocks, drive test out of their associated blocks, or be set into transparent mode allowing patterns generated elsewhere pass through the wrappers. Testbench wrappers are not part of the design, they are "compiled-out" prior to release. Setting internal wrappers to transparent mode creates a testbench for the SoC.

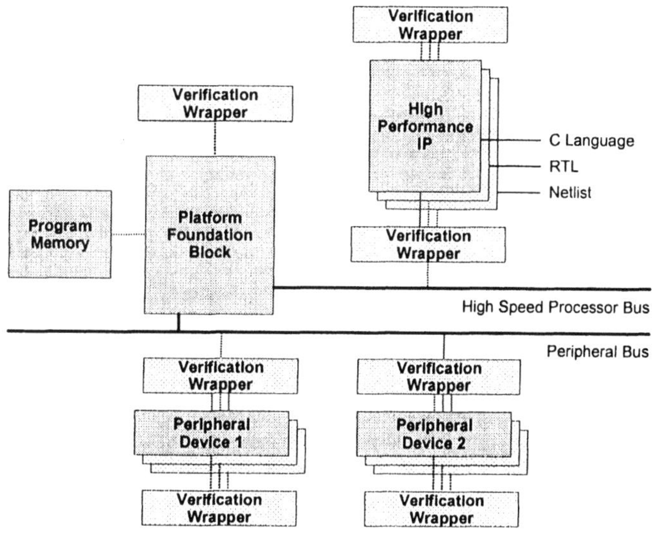

Figure 2-7. Functional Verification Example

The test wrapper functionality includes:
- Test vector translation- transforms test suites from one format to another allowing a single source to drive multiple views of the design. An example would be to transform frame based video images into a pin accurate serial bit stream.
- Test vector capture- captures a test suite for the block when an external source is driving. An example would be to capture a block level test suite from a system level test.
- Test vector playback- replays a vectors set that has been captured previously allowing blocks to be tested in isolation.
- Protocol checking- checks for bus protocol violations that appear on the wrappers bus interface.
- Results checking- checks the blocks responses to a test suite. This has two types of checking- expected response file and "golden" models co-simulation and compare.

- Code coverage analysis- quantifies the functional coverage of a test suite as applied to a specific block.
- Random pattern generation- Apply random patterns to the bus interfaces. The results are checked by protocol checkers or may also be checked with signature collectors.

In the platform-based approach to SoC, reuse of testbenches and verification models is a given, because the verification environment is first built for the platform, and then reused with little modification required for each derivative design. At this level, the only modifications needed are for any new IP blocks used, or for any blocks with modified functionality. Platform-based design allows the maximum amount of verification reuse when compared to a more custom, block-based SoC approach.

6.   HARDWARE IC DESIGN

In the SoC era, for both block-based and platform-based methods, the emphasis is on chip integration, not the individual customized design of each block. In the platform-based approach, it is typical to design the "hardware kernel" or "foundation block" of the platform as a pre-laid-out, customized and characterized reusable element which forms part of every derivative design. There are signs, however, that this pre-laid-out approach is slowly being de-emphasized. We are beginning to see that the value of reuse is shifting from the hardware implementation level to the system architecture level, where a well defined micro-architecture allows for the reuse of embedded software components and system level hardware models. This leads to a much higher degree of time savings compared to the time saved based only on a fixed hardware implementation. Irregardless of this change, the fixed hardware kernel (which may be "soft", i.e. at RTL) generally incorporates the processors, processor and system buses, memory interfaces, possibly some fixed portion of memory, and bridges to the peripheral bus. The platform-based design stage also usually fixes some part of the major SoC physical architectures such as timing, test, and power. In the derivative design process for a platform-based design, the hardware kernel is assembled together with new, modified or pre-characterized blocks from IP libraries to form the final hardware realization. Of course, the blocks to be assembled need to be designed in the context of the overall SoC design plan and constraints. The major steps [14] required in hardware design include:

- Front End Assessment (see Section 10)
- Chip planning

- Block design
  - RTL synthesis, place and route
  - Analog/mixed-signal block design
  - Memory block design
  - Processor core design and assembly
- Chip assembly
- Verification- timing, power, functional, and physical

Arguably, the most critical step is chip planning for both block-based and platform-based designs. In this chapter, we describe only this step. *Surviving the SoC Revolution* [1], Chapter 7 describes the entire hardware design task in more detail. Chapter 4, Section 3.3 (Philips) describes the SoC implementation approach used in the Nexperia Digital Video Platform. The overall chip plan dictates the ease of block design and assembly, and imposes the major constraints on timing, power, and test. Developments in hierarchical chip design tools offering advanced floorplanning, estimation and assembly capabilities for block-based approaches are fundamental for SoC. A depiction for these major steps is shown in Figure 2-8.

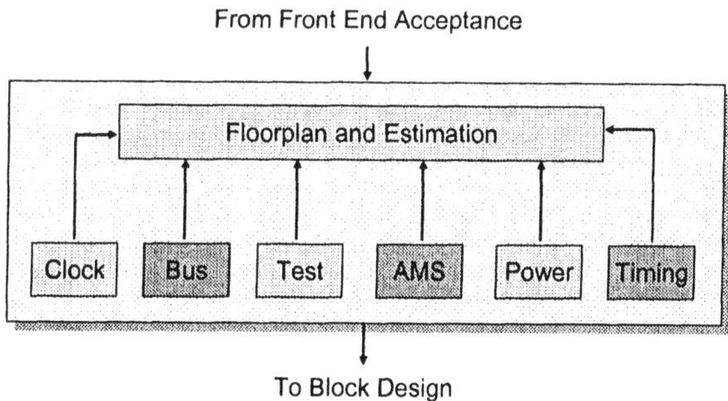

Figure 2-8. Primary Chip Planning Steps

In derivative design, the primary issue around clocks is how to attach them to the ones provided by the foundation block. This includes their distribution, their balancing, as well as how to handle asynchronous clocks. For the buses, the issue is to verify that they can be translated from a logical level to a physical level. For example, buffer interfaces may need to be inserted. Also "collars" may need to be wrapped around blocks to provide these buffers as well as translators from one bus protocol to another. Simple transformations may be required such as resolving "endian" issues. Design for test must also be considered up-front. For example, will a hierarchical or flat test approach be used? The IEEE 1149.1 standard also know as "JTAG" is almost always applied to modern designs. Deciding how each individual block will be tested- scan, built-in-self-test, and functional test is also critical. Finally, the tests need to be scheduled to minimize the time on the tester.

The inclusion of analog/mixed-signal (AMS) blocks poses additional challenges such as requiring additional "clean" power/ground lines (see Section 7). Often there are placement as well as routing constraints on these blocks to isolate them from the "noisy" digital blocks. The physical aspects of the power/supply lines also need to be considered. Will "power rings" be placed around each of the blocks or will the power/grid feed through the blocks? Power estimates will be required to estimate the power distribution needs for the SoC. Chapter 11, Section 2.2 (IBM) describes the issues when designing for minimizing power dissipation. Finally, timing needs to be considered. Hierarchical timing budgets need to be made. Clock balancing needs to be performed. Making life simpler or more difficult depending on the particular design methodology will be the fact that the foundation block may already be pre-hardened. Some of the constraints will already be characterized. Thus, the remainder of the chip needs to be designed "around" what has already been done.

Last but not least, is the floorplanning step itself. Once having taken all of the previous issues into account, the I/O pads need to be placed, the foundation blocks, as well as all of the additional blocks. Again, this is a challenge where some of the blocks will be hard, while others have yet to be physically implemented.

In chip planning, an incremental approach is often used where a designer is assigned full-time to be in charge at the chip level taking the SoC from chip planning to chip assembly. The designer continuously refines the chip-level view (e.g. via virtual silicon prototyping), as the design continues at the block level. The chip level routing and placement is repeated over and over as new data comes in to ensure that the overall chip is feasible and meets the system requirements.

Also as we move to 130nm and beyond, the process-technology related issues described in Chapter 11 (IBM), such as parametric variability, reliability, manufacturability, and affordability, become increasingly important to consider in the hardware IC design.

7.   ANALOG/MIXED-SIGNAL IN SOCS

Although an increasing number of SoCs are mixed-signal, by-and-large most of them are mostly digital with only a portion of the chip analog/mixed-signal (AMS). Many SoCs contain primarily digital devices with analog interfaces. The SoC design approach for AMS focuses on integrating AMS blocks into SoCs rather than authoring or packaging. A simple flow is shown in Figure 2-9.

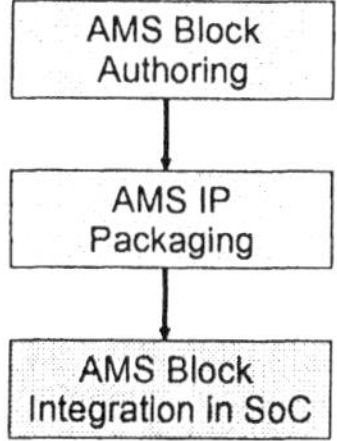

Figure 2-9. AMS IP in an SoC Design Methodology

The goal of authoring an AMS block is to design it for reuse as well as for ease of integration into the SoC. The idea of packaging is to assist in making integration possible by someone other than the IP author, i.e. by someone who is not an analog expert. One approach is to deliver the AMS block in such a way that the digital designer almost sees it as a "digital" block, thus allowing the integrator to use their standard digital design methodology to do the integration. The Virtual Socket Interface Alliance (VSIA) has done fundamental pioneering work in defining the standards for AMS block development and integration into SoCs [15]. The authoring and packaging aspects are described in detail in Chapter 8 of *Surviving the SoC Revolution* [1]. This section will focus on the integration issues.

Unfortunately AMS blocks cannot be packaged in such a way as to completely hide all of the additional issues associated with integrating an analog block compared with a digital block. When used, AMS components touch upon the entire hardware flow, and at different points in the design AMS issues must be taken into consideration.

If we take the major design steps as illustrated in Section 7, and use them as a guide to the steps that will be impacted, the first set of issues come into play with front-end acceptance. AMS blocks impact customer data validation, design feasibility assessment, and project planning and budgeting. The focus at this early stage in the design process is primarily to make certain that all of the data requirements are met for the AMS block. At the chip planning stage, all the components are impacted- clock, bus, power, test planning, floorplanning, I/O pad and block placement, as well as timing budget creation for use in block design. There are no simple rules as to what is to be done, and often an analog designer is required to assist with the AMS block integration. However, some basic guidelines include:

– Controlling substrate noise via placement

– Controlling noise around the periphery of an analog block via adding isolation trenches
– Avoiding routes over the analog block
– Controlling noise in the power rails
– Placing analog block far away from the noisy digital block
– Placing metal shielding completely around and over the analog blocks
– Controlling cross-talk noise within analog buses
– Limiting the length of the wire can deter the signal buses from attracting noise
– Controlling cross-talk noise in the I/O rings

Chapter 11, Section 2.3 (IBM) describes techniques to avoid signal integrity problems. Although this section is geared toward digital design, many of the techniques described can also be applied to analog.

In terms of block design, the only issue may be that an AMS block that is to be integrated has not been packaged properly. In this case a "collar" may need to be provided to handle some of the interface issues to the rest of the digital blocks. Application engineering support from the AMS block provider will be essential to making integration possible in this case.

The focus of chip assembly will be using much of the routing information provided with the packaged AMS block. This routing information includes electrical constraints such as capacitance, resistance, and inductance ranges that need to be met. Geometric constraints such as symmetry, and shielding constraints may also be provided. There may also be the identification of critical signals, required isolation from digital nets, and constraints for high frequency signal lines.

Finally, in terms of verification, as for timing, if the AMS block presents its digital interfaces with timing views (i.e. digital timing model for static verification and peripheral interconnect model [15]), then standard timing approaches can be used. Power information should also be provided with the AMS block. Functional verification can also be applied in the standard fashion if the AMS block provider has provided the appropriate models [15]. These include the functional/timing digital simulation model and a bus functional model. Physical verification should be straightforward. Design rule checking (DRC) exceptions (if applicable) should be listed in the packaging information. Finally, a netlist should be provided for layout vs. schematic checks (LVS).

The ease and reliability of AMS block integration has much to do with how well it is packaged. Today's AMS IP providers are realizing that a process for quickly designing IP (e.g. such as top-down design and synthesis [16]) is important to minimizing their costs, and meeting their time-to-market needs, but making sales is critically dependent on the right packaging and the ease of use of their IP.

8.   INFRASTRUCTURE

One unique piece of infrastructure for SoC design is an IP management system. Design information databases for IP blocks, platforms, and SoC designs continue to evolve. Reuse of design requires a well-architected and stable IP infrastructure in which design blocks (or platform aggregates of blocks) can be stored, searched for, found and reliably retrieved. Chapter 8, Section 3.1 (Altera) describes a typical set of IP in an IP library. But the database is only a foundation or substructure. Of equal importance are a set of IP management processes and procedures that allow effective use of the underlying database capabilities. Here, effective characterization and rapid search for IP blocks and platforms is a key requirement. An example IP management system is show in Figure 2-10.

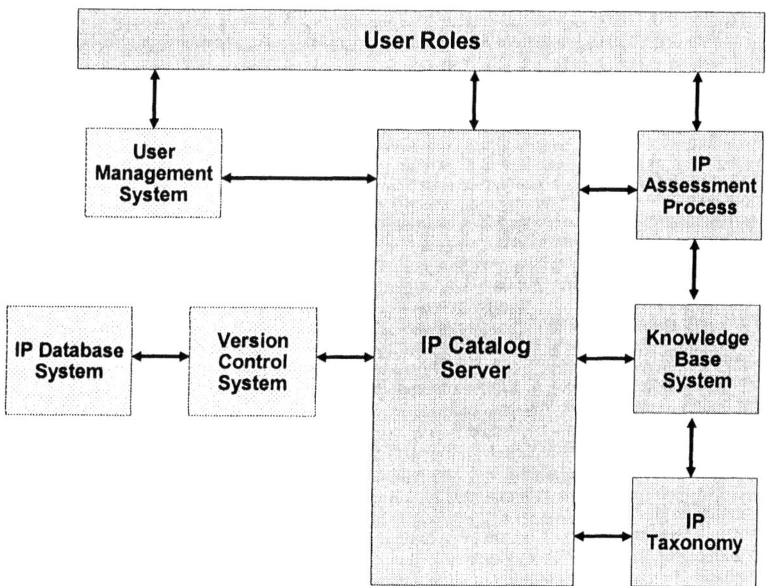

Figure 2-10. An IP Management System Architecture

The underlying components of such a system include:

- User roles which define who does what in maintaining and adding/retrieving IP in the system
- A version control system
- An IP taxonomy to assist in the placement and use of the IPs in the database.
- Knowledge base system
- IP catalog server
- IP assessment process
- User management system
- Configuration management system
- IP protection

The importance of this system is changing over time. As we move to more platform based approaches, the IP management system becomes less critical, but the number of non-differentiating IP (representing interface standards, basic building blocks) is formidable and must be managed.

9.   THE INTERFACES AND CROSS-DISCIPLINARY CHALLENGES

We have now described all of the major *design* tasks in the design methodology. This section emphasizes the need to also consider the interfaces between those tasks. Recent thinking in the area of SoC has developed the concept that what was considered to be separate domains of design where "over the wall" handoffs were sufficient, are no longer so.

Figure 2-11 takes a broader view of SoC and embedded system design to illustrate all of the design disciplines required. These include the system designer, printed circuit board designer, the package and chip I/O designer, the embedded software developer, the verification team, the IC design team (Analog, Digital, and RF), and the infrastructure team. Chapter 5, Section 3 (TI) describes their multi-disciplinary team for successful SoC platform design.

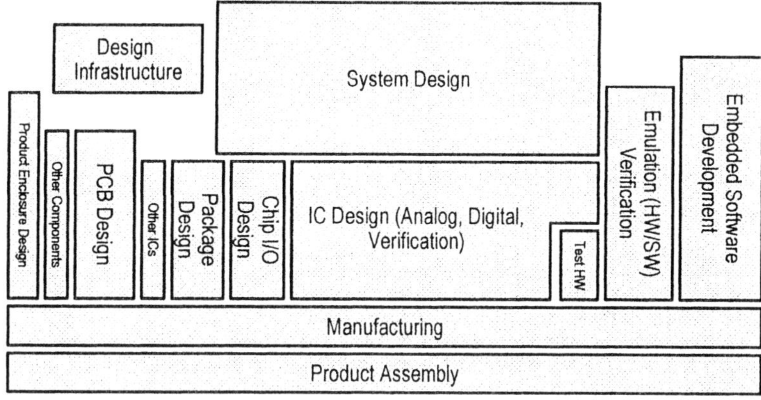

Figure 2-11. Broader View of Design Disciplines

Some of the key areas requiring interaction between the design teams are:

–   In Design- system design to IC design (analog and digital), IC design and chip I/O design, package design and printed circuit board design;
–   In Verification- timing verification from PCB to IC, timing verification between analog and digital, HW / SW verification methodologies; and
–   Links to manufacturing, links to test (analog and digital).

We are increasingly seeing these interface issues surface. Chapter 3 (Ron Wilson) talks about some of these interfaces in detail. Here, as an example, we describe the interaction between the package, the chip I/O and the printed circuit board designer. Although we talk about an SoC, not all systems will be integrated into a single chip. For cost and technology reasons, many will use multi-chip module (MCM), system in package (SiP) and even chip on board technologies to obtain the integration necessary. The design of wireless handsets is a clear example where the virtues of a single-chip vs. a two to three chip solution are being fiercely debated. Chapter 5 (TI) describes a platform for such a system. Although a chipset is not as dense as a single IC, it may be more cost effective and provide a faster turn around time. The use of flip chip may also become essential to this solution.

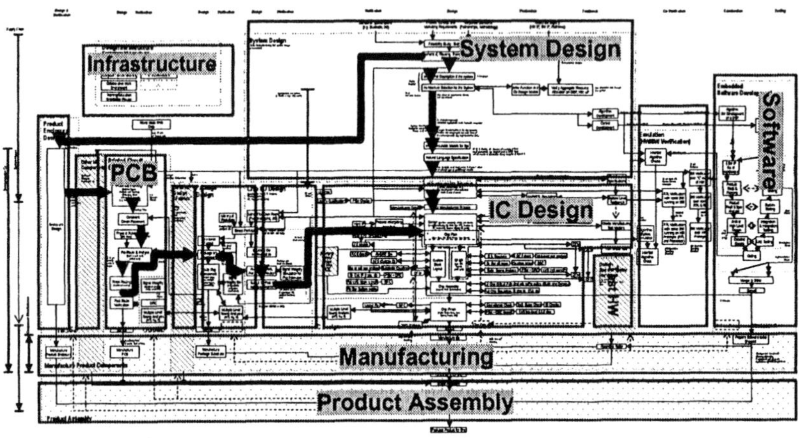

Figure 2-12. Loops Across Design Disciplines

The need to optimize I/O placement on the board, the package and on the chip simultaneously becomes critical. It is a type of three way co-design where designers require the ability to send constraints (e.g. pad locations, buffer locations, routing) between the different teams, and provide analysis capabilities across disciplines. These analysis capabilities include routability, timing and analog performance, power, crosstalk, and static verification of nets ("LVS"). Finally, because of the co-design nature, the solution needs to have a user interface useable by the board, the package, and the IC designer. The design must be able to start from any of these teams. This dependence is shown in Figure 2-12. Here we overlay the next level of design detail on the tasks shown in Figure 2-11. At this level, we can analyze the flow in more detail. Feedback loops become apparent such as this one that goes through system design, IC design, chip I/O design, packaging, and printed circuit board design. This is shown in the diagram by the black arrows.

Another link is the link to manufacturing. The handoff of just "gdsII" data is becoming increasingly insufficient to address all of the issues discussed in Chapter 11 (IBM). This is also discussed in Chapter 3, Section 7 (Ron Wilson). Chapter 3, Section 5 also describes the links beyond engineering, such as to marketing.

One solution to addressing these handoff issues is the adoption of industry standards. This is discussed in detail in Chapter 6, Section 2 (IBM) and Chapter 9, Section 3 (ARM).

10.   "META-METHODS"

Finally, we define a set of "meta-methods." Often we forget that not only do we have to go through all of the design steps, engineering is also about the management of data and communications between the engineers as well as with the end customer. We consider these as the design management steps that sit on top of the basic design methodology, hence the term "meta". Chapter 3 (Ron Wilson) touches upon the need to address this part of engineering as well.

These meta-methods [17] encompass a set of basic processes to reduce the risk of SoC design through the systematic collection and reuse of design experience. The processes include:

- Mechanisms for logging the design process, comprising of:
    - Metrics for measuring the design and design progress
    - Design sign-in points in the design flow
    - Efficient capturing of designer decisions and designer reasoning
    - Storage and retrieval of relevant design experience
- Certification to ensure completeness of the design process
- Qualification to ensure the sufficiency of design decisions
- Processes to use design experience for refining the design flow

Figure 2-13 shows how these steps are intertwined in the regular design flow [17]. Such processes are important in order to increase the confidence in taking on an SoC design project; to ensure effective reuse and minimal design time; to ensure that design experience is logged and used for systematic improvement of SoC design; and to allow quick assessment of the feasibility for a particular SoC design project.

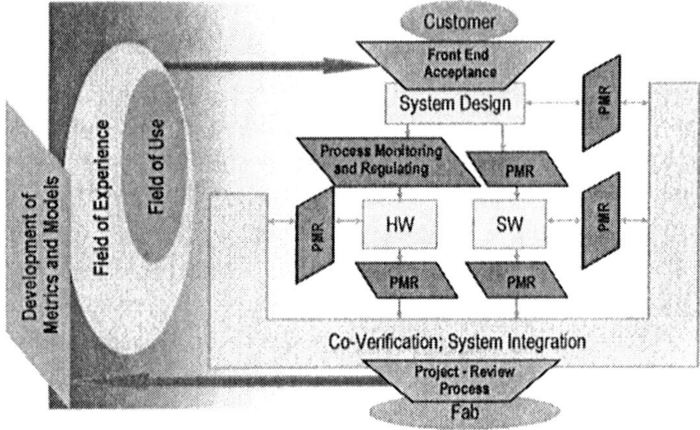

Figure 2-13. The "Meta-Methods"

The essential properties include:

- Monitoring- monitor and log progress defining the assessment points and criteria, and ensuring that the results are fully interpretable.
- Assessment- define the precision by which design decisions may be made.
- Communication- completeness of hand-offs, both forward and feedback.
- Adaptation- log repetitive issues in design flow and flag for restructuring.
- Refinement- reduce redundancies, clarify design checks and push design decisions to earlier in the design flow (if possible).

We will now describe five of the "meta" methods. Probably the most important is Front-End Acceptance (FEA). It is the process by which a design is "signed-in." It is characterized by *assessment* and *communication.* It provides a rigorous process for design-flow entry by ensuring that customer requirements are correctly interpreted. All information required as input into design flow is present. This also assists in giving the customer confidence that project-plan is realistic. This is essentially an assessment of design risk, and often requires actually designing part of the design (sometimes referred to as "dipping") to see if it can be done to schedule, and whether or not the design is feasible. Studying the platforms available is a typical step of FEA.

The complement to the check-in process is the Project Review Process. This is a "Sign-out" process. This can also be done at intermediate steps in the design flow. This is characterized by *assessment* and *refinement.* It checks that all information required of design is packaged to ship. There is formal completion of design tasks for a customer, a post-mortem of the design process, and adaptation for future design processes. If the design was simpler than expected, the Field of Use (FOU) may be extended. If the design was more difficult, recurrent errors are logged for future improvement.

The Field of Use (FOU) pre-defines the "scope" of where the "packaged" platform or IP block can be applied. It is characterized by *assessment*. It defines the conditions for the valid use of a platform (or IP). This includes valid absolute ranges of chip-level characteristics and valid correlations between design requirements. Thus, rather than having a customer try to make an independent assessment of whether or not a platform can be used in their situation, FOU attempts to predefine what these conditions are. Another way to view it is as an assessment of design risk relative to customer requirements, meaning that the further away the requirements are to the FOU, the riskier the design becomes.

Design experience is a very important aspect to design. Reusing IP blocks is an obvious way to reuse experience, but often the most effective form of reuse is to reuse designers as explained see Chapter 3, Section 6 (Ron Wilson). A meta method to try to capture this more formally is called The Field of Experience (FOE). This is characterized by *monitoring* and *assessment*. It is primarily implemented as a user interface to guide the logging of design experience. This includes a logging mechanism for quantified design experience, a statistical analysis engine, and a set of pre-determined data use-models. The system provides a structure for querying, storage of designer experience, a centralized location for accessing experience data. It also ensures that captured experience is complete. One example of data might be area or power estimates based on the number of gates in a design. This would serve as a complement to more analytic techniques to provide these estimates. FOE provides a mechanism for capturing prior design experience to provide this data.

Finally, the key to effective use of FOE is the development of metrics and models. This is a function of *refinement* and *assessment*. It provides the mathematical models supporting FOE queries. This step defines the set of useful metrics for measuring the FOE. It models abstract metrics in terms of detailed design characteristics, e.g. chip performance to bus/processor utilization. It also has the ability to detect lack of sufficient characterization of the FOE, and defines the concept of "design-closeness" for querying FOE.

Further details on some of these processes can be found in [17].

11.   CONCLUSION

The design of an SoC is a complex task requiring a myriad of design tasks and design know-how. A design team with an ever increasing mix of design disciplines must work together cooperatively to design modern SoCs. The basic steps have been described in this chapter, but by no means can one chapter provide both a detailed and comprehensive description of an SoC methodology. However, it is hoped that this chapter and the remainder of this book can provide a starting point from which one can derive one's own SoC Design Methodology to win in the SoC Revolution.

ACKNOWLEDGEMENTS

The basis for this chapter was a two and a half year effort at Cadence Design Systems in which a complete SoC Platform Based Design Methodology was developed. I would like to acknowledge the entire development team as well as the key architects on the team: Larry Cooke, Antoine Goujon, Merrill Hunt, Wuudian Ke, Christopher Lennard, Grant Martin, Peter Paterson, Khoan Truong, and Kumar Venkatramani.

REFERENCES

1.   H. Chang, L. Cooke, M. Hunt, G. Martin, A. McNelly, and L. Todd. *Surviving the SOC Revolution: A Guide to Platform-Based Design*, Kluwer Academic Publishers, Boston, 1999.
2.   G. Martin, "Productivity in VC Reuse: Linking SOC Platforms to Abstract Systems Design Methodology", *Proceedings of Forum on Design Languages: Virtual Components Design and Reuse*, (Lyon, August-Sept. 1999), p. 313. Also found in Chapter 3 of: R. Seepold and N. M. Madrid, editors, *Virtual Component Design and Reuse*, Kluwer Academic Publishers, Dordrecht, 2001.
3.   Martin, G. and B. Salefski, "Methodology and Technology for Design of Communications and Multimedia Products via System-Level IP Integration", *Design Automation and Test in Europe* (DATE) 1998 Designer Track, Paris, March 1998, pp. 11-18.
4.   F. Balarin, M. Chiodo, P. Giusto, H. Hsieh, A. Jurecska, L. Lavagno, C. Passerone, A. Sangiovanni-Vincentelli, E. Sentovich, K. Suzuki, and B. Tabbara, *Hardware-Software Co-Design of Embedded Systems: The POLIS Approach*, Kluwer Academic Publishers, Dordrecht, The Netherlands, 1997.
5.   F. Schirrmeister and G. Martin, "Platform-Based Design Helps meld EDA with Convergence Demands", *Wireless Systems Design*, p. 21, May 2000.
6.   G. Martin, and C. Lennard, "Improving Embedded SW Design and Integration for SOCs", *Custom Integrated Circuits Conference*, Orlando, May 2000, p. 101.
7.   H. Chang, et. al, Surviving the SOC Revolution, page 210.

8. K. Keutzer, S. Malik, A. R. Newton, J. Rabaey, and A. Sangiovanni-Vincentelli, "System-Level Design: Orthogonalization of Concerns and Platform-Based Design", *IEEE Transactions on CAD of ICs and Systems*, 19, 12, 1523, December 2000.

9. G. Martin, L. Lavagno, and J. Louis-Guerin, "Embedded UML: a merger of Real-time UML and co-design", *CODES 2001*, Copenhagen, April 2001, p. 23.

10. P. Rashinkar, P. Paterson, and L. Singh, *System-on-a-Chip Verification: Methodology and Techniques*, Kluwer Academic Publishers, Boston, 2000.

11. Janick Bergeron, *Writing Testbenches*, 3rd. Edition, Kluwer Academic Publishers, 2003.

12. L. Lev, R. Razdan, C. Tice, "It's About Time- Charting a Course for Unified Verification," *EE Design*, Jan. 28, 2003.   URL: http://www.eedesign.com/features/exclusive/OEG20030127S0055

13. F. Carbognani, C. Lennard, N. Ip, A. Cochrane, and P. Bates, "Qualifying Precision of Abstract SystemC Models Using the SystemC Verification Standard," *DATE 2003* Designer's Forum, Munich, March 2003.

14. Grant Martin, Henry Chang, "System-on-Chip Design: A Tutorial," *4th International Conference On ASIC Proceedings*, October 2001, pp. 12-17.

15. "Analog/Mixed-Signal VSI Extension Specification," *Virtual Socket Interface Alliance*, November 1999, web site: URL http://www.vsi.org/.

16. H. Chang, E. Charbon, U. Choudhury, A. Demir, E. Felt, E. Liu, E. Malavasi, A. Sangiovanni-Vincentelli and I. Vassiliou, *A Top-Down, Constraint-Driven Design Methodology for Analog Integrated Circuits*, Kluwer Academic Press, Boston, 1997.

17. C. Lennard, C. and E. Granata, "The Meta-Methods: Managing Design Risk during IP Selection and Integration", *The Intellectual Property System on Chip Conference*, (Edinburgh, November 1999), p. 285.

Chapter 3

NON-TECHNICAL ISSUES IN SOC DESIGN

Ron Wilson
EE Times

Abstract: It is not just technical challenges that confront designers of complex SoC devices. There are many organizational and management challenges as well. In this chapter we discuss structural, organizational, communications and other non-technical challenges and issues that SoC designers must face. These include the fundamental concepts of interfaces: between design groups, between hardware and software, between system designers and implementers, and between design and manufacturing. We conclude with a detailed discussion of foundry interfaces.

Key words: design interface, foundry, system design, IP

1. NON-TECHNICAL ISSUES INFLUENCE SUCCESS

The technical challenges of SoC design are widely discussed (in, for example, reference [1]). Less often is there discussion of another side of the SoC design challenge: the organizational requirements that a system-level IC imposes upon its design team. Certainly it is necessary for the SoC team, like any other design team, to take stock of the skills that will be required to complete the project, and to make sure that those skills are present in the team, or can be acquired.

But beyond that, structural issues assert themselves in SoC designs that occur infrequently in other types of chip design. These issues involve how the SoC design team is partitioned and how the subgroups communicate amongst themselves. And the issues arise not just from the size and complexity of SoC designs, but from the characteristics that distinguish system-level chips from other IC undertakings.

2.   THE UNIQUE SOC

There are at least three respects in which the SoC is distinct from other ICs. These are not just matters of scale, but qualitative differences between the processes of designing an SoC and a conventional chip.

The first issue is identified in the term "system-level." By definition an SoC is not just a big chip. It is a chip that contains the majority--or perhaps all--of the functionally important circuits in the system in which it will be used. That may sound like just a matter of scale. But in fact, the "systemness" of the SoC causes a profound change in the way the chip design team relates to other teams: most specifically the system design group, the verification team and the software team (Figure 3-1).

Note that in this chapter we will treat verification teams as distinct from design teams. This is for two reasons. First, it reflects reality for a significant number of management organizations. Second, it reflects the growing difference between design and verification groups in goals, vocabulary and tools.

The second major issue with SoC designs involves the diversity of functions that must be integrated into the chip. In all but the largest design teams, this diversity means that various blocks within the finished design will have come from different groups, some within and some outside the design team. Some of these groups will not be involved in the chip design process at all, and may not share a vocabulary, a tool environment, or even a language and culture with the primary chip design group.

Finally, we will address an issue that is old and familiar to some very powerful design teams, but that remains either an emerging issue or a lurking iceberg for many SoC designers. That is the need for foundry process engineers to communicate directly with the design team at a growing number of levels. This issue has hidden behind such catch-phrases as design for manufacturing, design-rule checking or simply good foundry relationships, but is in fact a significant organizational issue that must be explicitly considered by the design team.

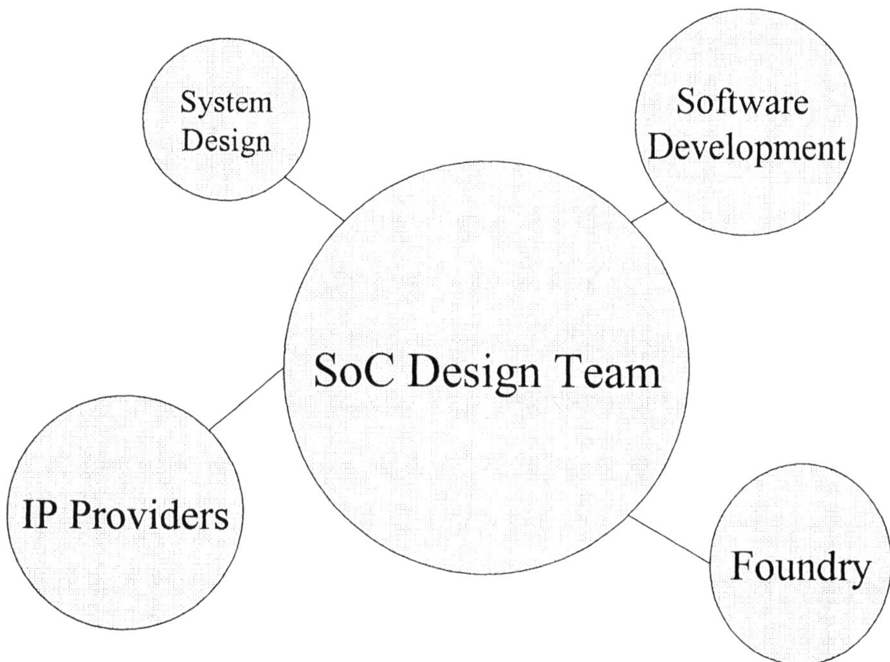

Figure 3-1. SoC design teams require a variety of external relationships.

3. THE CONCEPT OF INTERFACE

The central point of this chapter is that when information must flow between groups that are isolated from each other, whether by goals, methodologies, geography or culture, an interface is created. We use the word interface intentionally, because the analogy to an electronic interface is quite strong.

Many experienced managers, recognizing the need for large amounts of data to move between members of the design team, believe that the best topology for a design team is to have everyone literally in one room, the 'boiler-room'. With physical proximity and, more importantly, with the informal contacts that come from having coffee and lunch together, members of the design team build their own ad-hoc relationships, through which design data can flow very efficiently.

This topology has both advantages and disadvantages similar to those of other tessellated systems. On the plus side, no one needs to comprehend the entire structure of the design team or explicitly define interfaces between team members. Interfaces evolve naturally in response to the need for data.

They can also be quite dynamic, with some team members associating closely during early design phases and barely speaking later in the process, for example. On the minus side, relationships will only evolve when a member feels the need for information. If a designer doesn't know that someone is modifying the other side of a bus for which he is responsible, for example, he has no reason to talk to them about it. Further, information-sharing relationships, while they can be quite efficient, can be heavily influenced by personal relationships. Projects have come to grief simply because two designers who should have been sharing data hated each other's guts.

As projects grow to SoC proportions, these disadvantages begin to weigh against the advantages of the boiler-room topology. But a new factor arises that can render the approach completely intractable. As design complexity grows, design teams tend to fragment. Parts of the design group may be located at other facilities, or key designers may telecommute. Outsourcing may mean that critical parts of the team are in different companies, even on different continents. As we move toward SoC-sized designs, more formal interfaces between groups become more desirable and, eventually, mandatory.

While the interface metaphor is valuable in understanding these relationships, it should not be taken too far. People, after all, remain people, and will build their own ad-hoc networks no matter what the organization around them. Thus even if an interface between, for instance, a design team and a contracting firm exists explicitly on a corporate level as a contract, and implicitly at a working level as a set of Web documents that are continuously updated, there will also be key relationships between individuals. Often in emergencies it will be the individual relationships that function first and most effectively. Conversely, these relationship can be powerful obstructions to change. So the wise manager becomes aware of them and values them.

4.   AN INTERFACE DEFINITION

Using the analogy to electronic interfaces, it is possible to define an interface between two design groups rather precisely. Here is one way to do so.

An interface between design groups has three main attributes, each of which should be specified:
1. the data that is to be exchanged between the groups
2. the formats and protocols that govern the data transfer and interpretation
3. performance requirements to ensure that the interface can keep up with the project schedule

Each of these points is worth some discussion.

It is vital to specify the data that is to be interchanged between design groups. Yet in most projects this process is left almost to chance—in part because at first glance the specification appears intuitively obvious, and in part, ironically, because actually producing such a specification is so nearly impossible. But it is one of those instances of an impossible task in which added effort almost always brings rich reward.

To give some idea of the size of the task, consider the history of the Virtual Socket Interface Alliance (the VSIA [2],) an organization whose sole initial purpose was to define the content and format of the data that needed to flow from an intellectual property vendor to a design team using their core. Over the course of several years the VSIA grew from a seed to an organization with eleven working groups, hundreds of active technical contributors and a large Web site (www.vsi.org) full of documents. Even allowing for the self-propagating nature of organizations this gives some idea of the complexity of fully specifying even one data transfer.

Yet failure to undertake a definition is very dangerous. To return to the same example, the VSIA exists because, in part, of the horrendous experiences with failed interfaces between IP vendors and chip developers. So even a modest effort to agree at the outset on what will be the data needs of cooperating groups can reduce the chance of a project failure. The VSIA effort goes considerably beyond discussion, to an attempt to first establish a precise common vocabulary between IP providers and consumers, and then to create lists of deliverables and, in some cases, standards for evaluation.

4.1   Coherency across the Interface

That brings us to the second point, which might be summarized in the word coherency. It is of course necessary to make sure that each group in an interface is working with the same version of the data. Like any other interface, this sort needs a handshaking protocol. But because many of the interfaces in design groups are between teams from different disciplines, it is also vital that the data be interpreted the same way by groups on either side of the interface. For instance, when software designers work in parallel with the chip design team, both need a model of the behavior and software-accessible ports of the SoC. Their models are likely to be in quite different formats—one in Verilog and one in C, for example—but it is absolutely important that they be logically equivalent. This too is a problem with a less than perfect solution, but one that must be addressed.

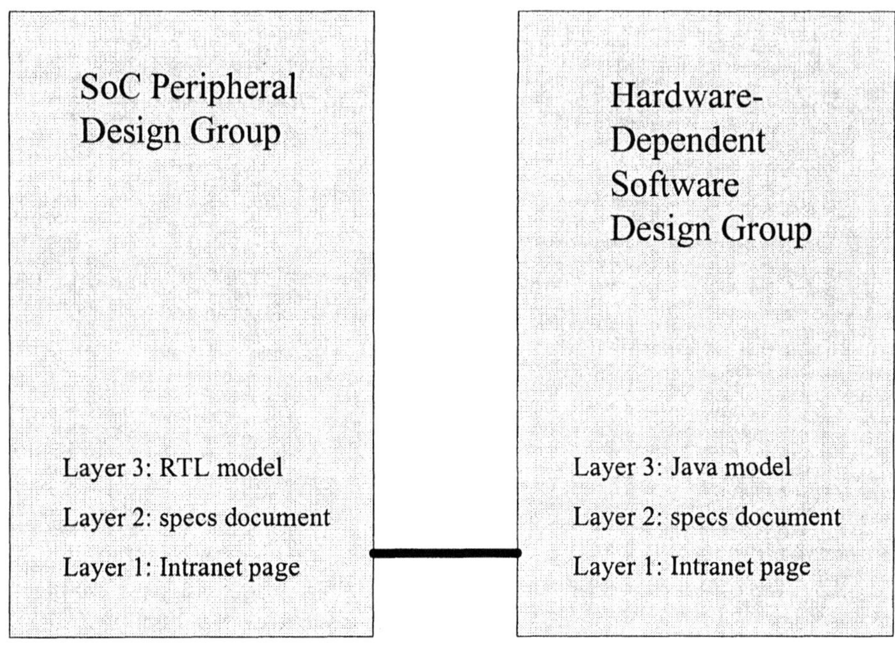

Figure 3-2. Making use of shared data may require a protocol stack in each of the groups that shares the data.

This issue can be thought of in terms of a hierarchical protocol stack (Figure 3-2). There is a physical layer, if you will, that involves physically transporting the data from one group to the other. Often these days that transport mechanism is the Web. There is something analogous to a media access and transport layer, in which messages are acknowledged, data is kept current, changes are detected and so forth. And there are higher layers, in which the data is translated into a form usable by the group. A change in the Verilog representation of a block by hardware designers, for example, may be need to be translated into a change in a Java transaction-level model of the block for use by software developers.

Finally, there is the matter of bandwidth. This is not just a matter of how many bytes of data can be moved between group A and group B in a second. That sort of bandwidth does occasionally become an issue, particularly in physical design. There have been instances where it became impossible to move the entire design data base between design groups electronically, and management resorted to cutting a tape and driving across town every evening to make sure all the groups had the current design. But raw data transmission speed is not the only issue here.

More important for most design teams is the time delay involved in getting the data into the right format. For instance, a system design group may control the requirements document that governs an SoC design. Despite recent inroads by system-level models in C or in systems-description languages, this document is still likely to be a huge binder full of text and block diagrams. If the system design team has a change of heart and modifies the requirements document, the design team's copy will have to be updated. But more important, the change will have to be translated manually into a C or RTL model for the design team, and into test bench code in e or Vera, or into assertions, for the verification team. The change hasn't really made it through the interface until it is in a form that can be used by the group on the other side of the interface.

The process of defining interfaces, then, is the process of specifying these three attributes--content, protocol and performance-- for each point at which data must flow between dissimilar groups within the design team. Depending on how complex the design is and how extensive the design team is, this process can become a significant design challenge in its own right. Like other major design efforts, reuse is vital: that is, success usually means employing as much as possible existing groups and relationships. Interfaces tend to get better with time.

Another interesting observation is that topologically, design groups often resemble the SoC architectures on which they are working. Very often an interface between major blocks on the chip corresponds to an interface between groups in the design team. On top of this must be added, of course, interfaces to verification, physical design and foundry groups that are not directly reflected in the chip topology. Interestingly, at the same time that the interfaces between groups in the design team are becoming more complex and less point-to-point, interconnect on SoCs is also moving to bus-oriented architectures, and peer-to-peer network-on-chip interconnect schemes are being investigated more widely [3]. To illustrate further, we will now examine some of the important interfaces in a typical SoC project.

5.   SYSTEM DESIGN INTERFACES

We would now like to consider some of the design team interfaces that are unique to the SoC design process. The first of these, what we will call system design interfaces (Figure 3-3), are made necessary by the very fact that an SoC is a system-level integration. Among many other implications, this means that the task of defining the SoC is on the order of—and in some cases nearly identical to—the task of defining the system itself.

In conventional IC design, the problem of requirements definition is well bounded. Often, the entire function of the chip lies within an industry-standard specification, such as the Peripheral Component Interface (PCI) spec. Or the function is commonly understood by members of the design team and their customers, so that little more than definition of I/O pins and clearing up a few options is necessary. This might be the case for a functionally simple device such as a serializer/deserializer (SerDes) block. So little energy goes into constructing or interpreting the definition of the chip.

For a system-level IC, however, the situation is quite different. The behavior of the chip may well be more complex than the behavior of the system in which it will be used. And that system behavior, lying in the realm of system design rather than digital design, may not be well characterized, or even entirely thought out.

This means that creating a model of the behavior and port structure of the SoC with sufficient precision to permit a design to begin may require considerable interaction with system designers, and even with other, non-design groups such as marketing. It is vital that the model the design team is implementing be close to the model system design is assuming. But it is far from given.

This would be complex enough. But in the SoC design there are other clients as well who must understand the behavior of the chip in some detail. Most prominent among these is software development. The majority of SoC designs will be done in parallel with a software development effort—often a very large one. Obviously it is vital that the software team have a black-box model of the chip's ports and behavior that is congruent with the chip design team's white-box model. Again, this is far from automatic. As understanding of the interface between chip design and software development improves, we are seeing a proliferation of models just for use by software developers. These range from simple behavioral models to models that give increasing detail: instruction-set simulations of processor blocks, transaction-level or cycle-accurate models of ports, and eventually models of sufficient complexity to indicate, for example, when a block is going to stop accepting data because its input buffer is full.

Finally there is, increasingly, a third participant at this level. More and more SoC projects are separating verification teams out of the design team, so that they form a separate group not directly in contact with the designers. But the verification team must rely on the formal definition of the chip in order to construct test benches, write assertions and plan verification strategy.

So the system design interface is a four-port interface, connecting the chip design team with system design, software development and verification groups. Following our model, we should attempt to identify the data that must flow through this interface, the protocol used and the performance.

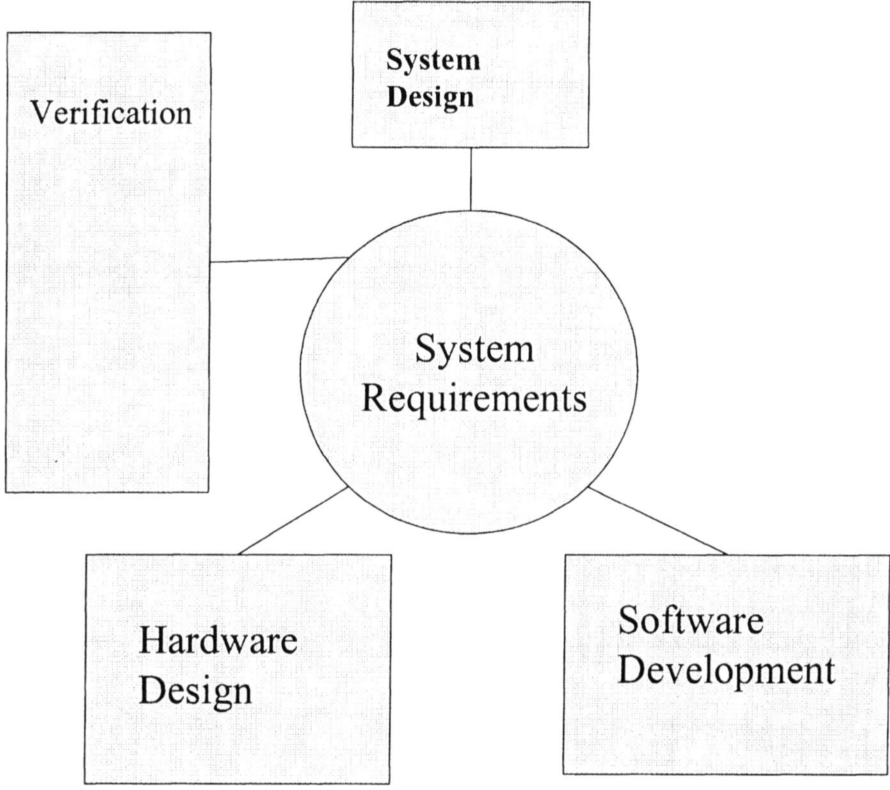

Figure 3-3. The System Design interface.

The nature of the data may be stated concisely, if not precisely. The interface must communicate to all of its ports a model that describes the I/O structure and behavior of the SoC. Clearly, this data is presented differently at each port.

The system designers need a black-box model of the SoC that accurately represents its function, external timing and I/O configuration at a port level. For system purposes it is often sufficient to think of the power connections as just another port. Application software developers, in contrast, generally try to avoid having to care about power, and may need only a general idea of timing. Driver developers may need a bit-accurate model of the I/O ports accessible to the software, and a similarly bit-accurate model of the chip's functionality. Verification engineers, for their part, need essentially the union of the two previous models. In order to verify the functionality and timing of the chip they must have a bit-accurate, cycle-accurate model.

5.1   System Design Interface Protocols

The protocol necessary to create these models, and to keep all the different views of the SoC coherent through the changes that occur at all the four ports during development, is one of the great unsolved problems of design management. In many design groups the data are contained in a collection of text documents and block diagrams—a technique dating from the earliest days of electronic design. The text approach has many advantages: it is intuitive, requires no special tools, and is easy to update.

But there are problems as well. Since each group has different data needs, one master requirements document tends to evolve into a network of a master document and four more specialized documents—one oriented to the needs of each client. This creates the problem of coherency—that is, to be certain that each group's document is consistent with the others. Another problem is sheer intractability. A recent framer IC design at Agere, for example, roughly an 11 million gate chip, required not only a substantial black-box requirements document, but an engineering specification of no less than a thousand pages. Obviously with documents of this size forming an accurate abstract view of the chip is unlikely, and revision control—even with software assistance—is a formidable problem.

Numerous methods are used to resolve these problems. The most common and oldest approach is the design review meeting. These meetings may be held at major milestones, at regular intervals, or any time a change is proposed to the documents. Attendees generally include managers of individual teams and any additional technical staff deemed necessary. Managing such a meeting so as to produce a definite result in a finite time is an art not to be underestimated. But well managed, review meetings can be extremely effective at identifying ambiguities in a spec and ensuring that the various documents remain consistent.

A more ambitious approach is the technique of repeated builds. Some design teams have had considerable success by creating a complete bottom-up build of the chip at regular intervals from the current design data. That build is then subjected to regression tests that have the effect of validating the current state of the design against the requirements—at least, against the requirements as understood by the verification team. If the software development team and system designers follow similar disciplines at appropriate intervals, it is likely that any discrepancies between the various views of the SoC will show up within one build cycle. The amount of work necessary is enormous, but the majority of it can be automated, so that once established, an iterative-build methodology does not impose huge demands on the designers. The process has significant value for coherency within the design team and for design convergence as well.

Another technique may prove useful in systems in which the major functional blocks are loosely coupled. That is decomposition. In effect, the various clients cooperate to disassemble the system block diagram into pieces that are small enough to be manageable by conventional, non-SoC methods. This techniques works particularly well in cases where the SoC comprises a microprocessor core, memory and a group of simple peripherals, all interconnected by an industry-standard bus. All four system design interface clients can take the SoC apart into its basic blocks by simply pulling the blocks off of the bus. All of the blocks are then readily handled in isolation, and the interactions between them are constrained by the bus protocol.

All of these solutions are somewhat ad-hoc. Many design managers have wished for a single language in which the system requirements could be expressed in a form that is both executable—to meet the needs of system designers, software developers and verification engineers—and synthesizable, to create a verifiable, automated path from specification to implementation. Unfortunately, such a technology has remained, to date, beyond reach. The nearest approach has been to create a model of the SoC requirements in a programming language, often C++ or Java. This provides an executable model that is of direct use to the systems designers and software developers, and is of considerable help to the verification team. But there is no direct path either to verify equivalence between such a model and an RTL model, or to synthesize RTL from the programming language code [4]. Yet the utility of such models is so great that many—perhaps most—SoCs today begin life as a behavioral model in a programming language.

Recently a number of so-called system-level languages have been developed. Some of these are elaborations of C++ to include temporal concepts and some sense of hardware structure—SystemC for example. Others are extensions in the opposite direction—elaborations of RTL languages to provide systems description constructs, such as SystemVerilog. Others are more complex exercises, starting from a clean sheet of paper to create a systems definition language. While there has been considerable interest in these approaches, it is too early to judge the long-term prospects for any of them. At the moment SystemC appears to have both broader use and the backing of key tool vendors.

5.2   System Design Interface Performance

Finally, the performance of the system design interface must be considered. The rapidity with which a change can move from proposal to inclusion in the documents is generally not an issue. In fact, if it becomes an issue that is in itself usually an indication of a problem in one of the client groups. The SoC requirements shouldn't be particularly dynamic.

But there is a more important consideration. The real delay in the system design interface is the time it takes a change from one group to work its way through to the other groups' models and be reflected in their work. This is where the traditional text-based technique gets into trouble. Even with good revision control and notification, it can take days or weeks for the appropriate person in a group to get the time, read the change and understand its full significance. Executable models coupled with a regular system build and verification cycle can be very valuable in overcoming this problem, as can, clearly, appropriate decomposition of the SoC.

6.   THE IP INTERFACE

The most readily discussed characteristic of system-level IC design is its reliance on intellectual property from sources outside the design team. Most designers agree that external IP is necessary to designs at system-level. It has been less often remarked that the use of IP creates the need for another type of design team interface.

Whether the IP comes from a previous design, from an IP vendor outside the company, or even from a different design group working in parallel with the main chip design team, the conditions that necessitate an interface are present. Groups separated by distance or discipline need to communicate with each other. So we can once again apply our interface definition, and discuss the data, protocol and performance of the IP interface.

6.1   IP Interface Data

The IP interface presents an interesting contrast to the system design interface. In the latter, it is relatively easy to define the data that must move across the interface. In the case of IP, however, just defining the data has turned out to be an enormous job. As previously mentioned, whole organizations, such as the VSIA, have devoted years to the task. It is easy to generalize that the IP provider needs to communicate to the chip design team all the information that is necessary to successfully design the IP into the chip. But specifics become elusive quickly. As one gets deeper into the design process new types of necessary data keep emerging. The partial list of VSIA Development Working Groups (DWGs) in Table 3-1 will perhaps give some idea of the scope of the issue.

Table 3-1. Some Currently active VSIA Development Working Groups

Group	Charter
Analog/Mixed-signal	analog IP documentation, signal integrity
Functional Verification	functional verification guidelines
Implementation Verification	data standards for implementing, verifying IP
Manufacturing-Related Test	testability of IP
On-Chip Bus	interconnection of IP blocks
System-Level Design	system-level views of IP
Virtual Component Quality	assessing quality of IP deliverables
Virtual Component Transfer	documentation and transferability of IP

In abstract terms, the problem is that all of the data files that an IP vendor passes to a client describe attributes of the implementation, but none of them unambiguously describes the designer's intent. RTL may describe the register-level topology of a netlist, but it does not contain the synthesis directives, timing requirements, test bench cases and other types of information necessary to ensure that the resulting block on the die behaves as the original designer expected.

More concrete examples might help. If an IP block is provided as synthesizable RTL, it must be accompanied by enough information to permit the user to issue the right synthesis directives to get what he intended. There must also be enough information about the intended behavior of the block to permit the verification team to do their job. Enough timing information must be there to ensure that the resulting block will function correctly.

To dig a bit into the latter example, typically an RTL design will be accompanied by timing files. But those timing files can be remarkably ambiguous. The original designer of the RTL may have worked out the maximum acceptable timing for each path in the block. Or, he may have simply kept tightening the constraints on critical paths until the synthesis tool produced a block that met specifications in his particular environment. The block may be substantially over-constrained for other situations. And those constraints may lead to absurd behavior by the synthesis and physical design tools. But in order to relax the constraints, the user may need to understand the theory of operation of the block at a level that nearly defeats the purpose of using IP in the first place.

Similar problems occur in other areas of IP implementation, including power, clock and test insertion, physical design and, throughout the process, verification.

A common thread to the problem of defining the data is that the true scope of the design team's data needs may not become clear until the design is already underway. And then a request for additional information or for help may bring the project to a screeching halt until the right person from the

IP provider's team can be put in touch with the right person from the design team.

6.2   IP interface protocol

This uncertainty influences the protocol necessary to conduct the business of the IP interface. Unlike the system design interface, the data needs of the IP interface do not lend themselves to a static delivery mechanism such as a requirements definition document. The data that are being exchanged, after all, are not about an end objective, but about the means of getting there.

Most design managers feel that there is no substitute for person-to-person contact between the IP creators and the IP consumers. The guy who wrote the code can tell you what he was trying to do in a direct way that no collection of data files can accomplish. And in fact most design teams who report success with complex IP in advanced processes say they enjoyed close working relationships with technical people from the IP providers.

Yet it is not feasible in most cases for members of the SoC team to have free access to members of the IP provider team. The IP provider is usually supporting a number of clients at once, and the SoC team may well be dealing with a number of IP providers at once. Communications would become untenable just when they became the most necessary.

This leads most design teams to assign a single point of contact on each side of the interface—usually a customer support engineer on the IP side and an engineering manager on the SoC side. These two individuals can then both monitor the flow of planned data, track simple questions directly and establish temporary connections between team members as necessary to solve particularly difficult problems. In some cases these latter contacts end up with team members from one side moving into the other side's facility for extended lengths of time.

We need to look at two special cases in which the above generalization may not be helpful: "silent" IP providers and IP providers who use a technology other than digital logic. In these cases special care must be taken.

First, there are cases where the IP provider is not going to provide much assistance during the design project. This rarely occurs in the case of commercial IP: pretty much all the surviving commercial IP providers understand the need for person-to-person contact between creators and consumers. One case in which this may still be an issue is when an IP vendor views the IP as self-evident—that an informed engineer familiar with the class of IP would herself understand everything necessary to employ it. An example would be PCI interface IP, which has gradually passed from black art to intuitively obvious in the last few years. The assumption of obviousness may be quite correct for some SoC teams, and quite wrong for others.

Table 3-2. Some instances of silent IP providers

Provider type	Behavior
Commodity IP source, i.e. public domain	Support available only from other users
Archived internal IP	Support only from data in the archive
IP from different technical discipline	Data and requirements need prior understanding

But it is a particular problem, ironically, with in-house IP for which the original design team has dispersed. In this case there is no existing body of knowledge the IP user can turn to. If the original designers are still with the company, they have long since gone on to other projects.

It is in recognition of this problem that many large organizations, among them Texas Instruments, Motorola and Fujitsu, have undertaken large programs for IP reuse [5]. The goal of these efforts is to capture, before the IP creation team has dispersed, all the knowledge that might be necessary to reuse the IP. It is this attempt to predict *a priori* what data will be necessary that leads to massive data collections such as those described by the VSIA working groups.

It can only be said that the jury is out on these efforts. There have been clear successes in reusing even quite complex IP. But these successes have required deep cultural change by the engineering organizations that created the IP, created elaborate data structures, and may still prove in the long run only to be effective when there is some access to the original designers in cases of crisis.

The other problem arises when the design group that creates the IP is from a different branch of the electronics discipline than is the SoC team. Obvious examples are advanced memory and analog IP. In this case direct contact between team members across the interface may not prove all that productive, as there may not be a common language. It can, for example, be a mutually frustrating experience for an RTL synthesis expert and an analog circuit designer to attempt communications. In this case mediation by technical management is essential. But it is equally important for the SoC team and the specialized IP team to set out at the beginning of the project what information needs each group anticipates, creating a template to be filled in as the design progresses, and monitored by an appropriate manager. Models at the appropriate level of abstraction can also be valuable—for example, a C++ functional model of an analog block can be a big help to the digital design team. This won't avoid all the problems, but it will head off many.

Performance of the IP design interface is a question that must be dealt with by design management on a case by case basis. We would suggest a triage approach. Some external IP, it will be evident, is either so obvious or so familiar that it will simply blend into the design flow along with newly-created data. For this IP only an emergency contact need be established, and it will probably never be used.

For other IP, there will certainly be issues during the implementation of the SoC. The design manager should first, ensure the ability of the IP provider to respond to requests—or convince himself of the adequacy of the archived information that supports the IP in the case where the provider is no longer available. Then some individual on the design management team should be assigned to develop a personal contact with the provider, and to serve as a conduit of information on the IP for the duration of the design. This person should be the focus for all communications to the provider, should track questions and responses and should not hesitate, if an IP issue becomes a critical path, to call in senior management.

If all this sounds like a great deal of work, it can only be admitted that it is. Few design managers explicitly identify IP interfaces at the beginning of a project and intentionally design them for adequate performance. Sometimes the interfaces are allowed to develop by evolution, and often they are erected ad-hoc in the face of an emergency.

It has been observed that successful design teams tend to stay together, even moving as a group from one company to another, much like journeymen in the age of craft guilds. It has also been observed that these design teams tend to maintain the same IP vendors from job to job, preserving interfaces that have evolved through trial and trouble and that have come to be trusted. There may be a message there.

7.   CONCLUSION: THE FOUNDRY INTERFACE AND BEYOND

It has always been necessary for the first design teams to work on a new process to work closely with the process integration staff of the fab. This has been just as true in the era of independent foundries as it was in the early days of integrated device manufacturers (IDMs) —chip companies who owned their own fabs. In fact, it has become common for independent foundries to form development partnerships with a few select design teams as a process is moving into pilot phase, and for the two to work together to bring the process into production. Often these relationships can be every bit as close as the relationships between chip design and process integration teams within an IDM.

It has also always been common for design teams who pushed the performance envelope to work closely with process engineers. For these teams, design rules were made to be broken, and process limitations were there to be explored, not worshiped. Once again, chip design and process engineering groups from the two companies worked closely together.

For stable processes and conservative designs, however, the interface between chip design team and foundry has been a carefully documented hand-off, based on strict lists of deliverables and automatically checkable design rules. This has been one of the most explicitly recognized and carefully documented interfaces in the entire SoC flow.

7.1 The Foundry Interface

The interface opens with the foundry passing a process design kit (PDK) to the design team. This contains the libraries, design rules, lists of deliverables and guidelines necessary, in theory, for a competent design team to hand off a tape to the mask shop with every confidence that they will get back exactly what they asked for. And, for stable processes, that has generally been the case. Problems that cause respins are almost always due to asking for what one does not want, not problems in getting what one requested.

But a number of factors are beginning to call this interface into question, even for stable processes. One factor is the rise of sub-wavelength imaging. Starting at 130 or 90 nm, depending on the process, some of the critical mask layers require optical proximity correction features, phase-shift plates or both. In 130 nm processes, for the most part, the existence of these features on the mask could be hidden from the physical design team by keeping the features inside the cells, and perhaps by adding some spacing or orientation rules to prevent phase shift areas or optical proximity correction (OPC) features in adjacent cells from interfering with each other. This approach would not work in interconnect layers, but the interconnect masks rarely used such features.

As processes grow more demanding, mask makers may no longer be able to hide this added complexity from physical designers. If cell designs are to be kept anywhere near optimum, they will have to take into account what is in the nearby cells. This is not limited to merely adjacent cells, because some OPC features can have effects more than one cell-width away.

So this growing complexity may have to be passed to the chip design team through an enormous increase in complexity in either cell selection— one estimate holds that a basic cell library would have to expand by an order of magnitude to include all the possible combinations of cell designs to resolve adjacent interference problems—or an enormous increase in design rule complexity. If the tools are unable to cope with this growth in complexity—and today it appears that they initially will not cope—there will have to be a direct human-to-human interface between at least the physical design team and the mask shop.

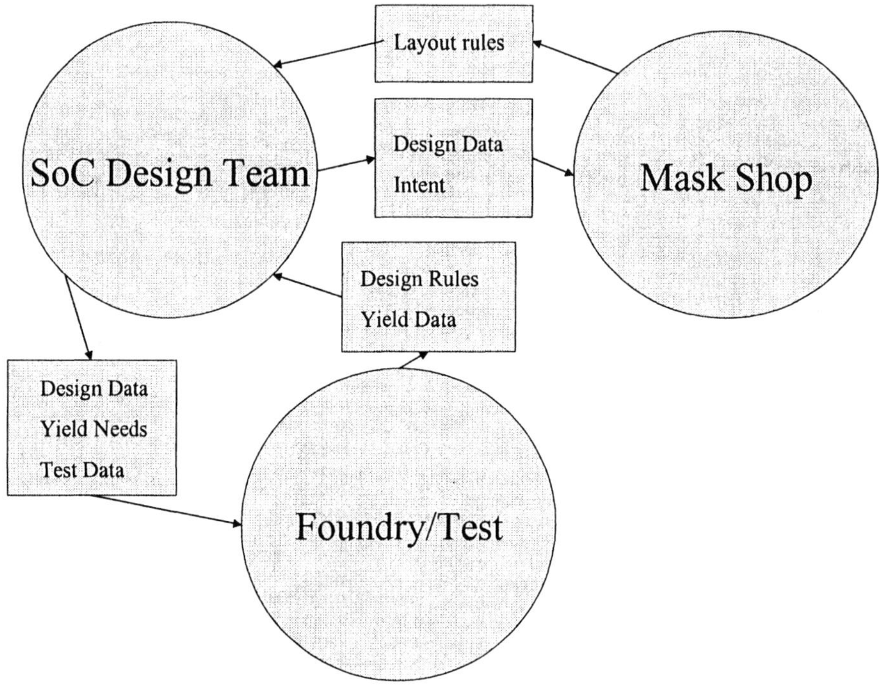

Figure 3-4. A wide variety of data may flow between design team, foundry and mask shop.

This interface may be necessary for other reasons as well. Mask inspection is becoming difficult at least as rapidly as is physical design. It is no longer feasible to inspect a mask and reject it if it differs from the ideal. The mask maker has to ask if the differences are going to produce a failure on the wafer, or if they can be ignored. Increasingly, that question can only be answered in light of the original design intent. Once again, there needs to be an interface between mask shop and design team. But this time the RTL designers may be involved, not just the physical design crew.

Of equal concern is the process itself. Traditionally, if the process failed to reproduce the features as described in the mask set, it was considered a failure. But, gradually and quietly, this standard has been eroding. Today, there are areas in which the correlation between features on the mask and features on the wafer is purely statistical. One example would be vias. In even 130 nm processes, the successful formation of a via is a function of the surrounding metal topology. Under some circumstances, such as in the middle of a wide metal line, the process rules do not guarantee that any given via will form—they only give the overall probability that a particular via will form. This is due to a number of failure modes, but primarily the stress that the large piece of metal exerts on the bottom of the via as it expands and contracts with temperature, during both processing and actual operation.

Foundries have overcome this problem by specifying circumstances in which redundant vias must be designed into the layout. These have no logical or even electrical function other than to increase the odds that at least one of the vias will survive processing and thermal cycling.

Other such examples are proliferating. It is necessary to insert dummy metal in order to keep the average ratio of metal area to dielectric area in a small region relatively constant across the wafer. Otherwise planarization can result in removal of too much copper, leaving interconnect lines thinned, increasing resistance and risking mechanical failure. And as processes make more aggressive use of OPC and phase shifting, the shape of the features actually formed on the wafer becomes less and less exactly predictable from the features on the mask.

Foundry engineers are attempting to keep all of these issues contained within the design rules. But it is increasingly likely that issues will escape from time to time, requiring a direct interface between physical designers and process integration teams even in stable designs.

7.2   Emerging Interface Requirements

In this increasingly statistical environment, another whole facet of chip design is emerging [6]. To design for speed, power, and test we now must add design for yield. This new requirement asks that the design team tune the design—often from the very beginning of the process in planning and RTL coding—to improve the odds that the process will yield well and that the dice will continue working in the field. As of now tool support for this undertaking is sparse and immature, but it is definitely emerging. This issue, too, will require that representatives of the design team from all parts of the flow have an interface to process engineers, at least for the near future, until the information that the process people know about yield can be encoded in tools.

Finally, there is the matter of failure analysis. In most of the electronics industry, the fate of a failed IC is simple. Once the chip is identified, it goes into the nearest waste basket. But in the automotive electronics segment, OEMs frequently demand rigorous inspection of failed chips, isolation of failure mechanisms and reporting on remedial actions. As processes become more statistical, yields more variable and field failure mechanisms more persistent, this concern about failure analysis is likely to spread.

This trend, should it emerge, will require yet another interface—between design and failure analysis teams. In IDMs who work on the edge, this interface already exists. Designers are accustomed to being called to participate in Focused Ion Beam, IR imaging or electron micrograph analysis of their chips. But for many SoC designers, this will be a new experience. Like the other interfaces we have discussed, it will require communications between groups with highly dissimilar vocabularies but a common concern.

7.3   Conclusion

This discussion has made some attempt to identify the important interfaces between members of the design team and other groups with whom they must share information. By callously appropriating the notion of data, protocol and performance from the world of electrical interfaces, we have attempted to give designers and managers a framework for planning and evaluating these interfaces. In the best of worlds, designs could be planned and design teams organized to require the minimum number of interfaces. These interfaces could then be explicitly recognized for the potential weak links that they are, and explicitly designed rather than simply allowed to develop. The impact on overall design success would be significant.

REFERENCES

1. Henry Chang, Larry Cooke, Merrill Hunt, Grant Martin, Andrew McNelly and Lee Todd, *Surviving the SoC Revolution: A Guide to Platform-Based Design*, Kluwer Academic Publishers, November 1999.
2. The Virtual Socket Interface Alliance (VSIA) technical documents and specifications represent a compendium of thinking and negotiation on the data required to reuse IP. The documents are available through URL: http://www.vsi.org/library/specs/summary.htm/.
3. William J. Dally, and Brian Towles, "Route Packets, not Wires: On-Chip Interconnect Networks", *Proceedings of the Design Automation Conference (DAC) 2001*, pp. 684-689.
4. W. Stoye, N.Richards, D. Greaves, and J. Green, "Using VTOC for Large SoC Concurrent Engineering: A Real-World Case Study", *Proceedings of DesignCon 2003*. (Example of the challenge of coherency between RTL and C models).
5. Michael Keating, and Pierre Bricaud, *Reuse methodology Manual for System-on-a-Chip Designs*, 3rd edition, Kluwer Academic Publishers, June 2002.   Available at ftp://cva.stanford.edu/pub/publications/onchip_dac01.pdf/.
6. Mark Lavin, and Lars Liebmann, "CAD Computation for Manufacturability: Can We Save VLSI Technology from Itself?", *Proceedings of ICCAD 200, pp.424-431*. (An excellent overview of issues at the design/manufacturing interface).

Chapter 4

THE PHILIPS NEXPERIA DIGITAL VIDEO PLATFORM

J. Augusto de Oliveira and Hans van Antwerpen
Philips

Abstract: This chapter will outline the challenges in platform development for digital consumer home devices delivering multimedia content and novel services and applications. It will detail the requirements on this market and the approaches taken by Philips Semiconductors when developing the Nexperia-Digital Video Platform. The Nexperia-Digital Video Platform (Nexperia-DVP) comprises a family of Systems on a Chip (SoCs) and a software platform that allows Philips' customers to build cost effective, flexible Digital Video appliances.

Key words: Platform, Nexperia, SoC, video, digital video platform, consumer electronics

1. NEXPERIA DIGITAL VIDEO PLATFORM

1.1 The Digital Video Revolution

The transition from Analog to Digital Video is transforming the way we enjoy home entertainment: in addition to the higher quality of the video and audio programs, we are also experiencing novel ways to navigate, store, retrieve and share the digital programs as well as access to new interactive services and connectivity possibilities.

Our home entertainment systems will be implemented with a number of Digital Video appliances, such as Digital Televisions (DTVs), DVD Players, Digital Video Recorders and Set-top Boxes. These home entertainment systems will *connect* to each other and to the productivity cluster around the PC and to mobile devices like cellphones and automobiles via wired/wireless networks and/or removable optical and solid state memories.

In comparison with their PAL and NTSC analog predecessors, the new Digital Video appliances, in addition to being able to decode program streams in a compressed digital format, will also include the computing power to navigate and process the digital stream. For example, a DVD Player, besides decoding the digital video, also provides the user with a simple yet sophisticated navigation system; and DTVs export Electronic Program Guides, greatly helping the consumer with the navigation of the live programming and the possible time-shifted recorded content.

Philips Semiconductors is developing a range of solutions for digital video appliances [1], based on a vision of this technology that is both imaginative and firmly rooted in real needs. The Digital Video vision must deliver valuable practical benefit to consumers. Digital Video appliances will provide consumers with access to and interaction with a powerful, coherent home network via an easy-to-use interface, which allows them to concentrate on what they want to do: to access their content-op-choice any time, any place with their device of choice. Users will focus on activities and needs, simply using the most convenient appliance, whether it's a DTV, a PC screen or any other part of the network.

1.2   The Philips Nexperia Platform Approach

Philips Semiconductors decided to serve the application domains of digital video and mobile with a *platform approach* that we call Nexperia [2]. The motivations for the Nexperia platform approach can be clustered around two main points: *rising product complexity*, both in silicon and software; and *commonality of functions* for the Digital Video scope, encompassing the following trends:
- Continued demand from end users for products that are *simple to use* but implement new applications and services, often leading to *internal product complexity*. Here we distinguish the internal complexity of the product, which makes new features possible, from the external complexity [3] that determines if the product is easy or difficult to use.
- Relentless increase in the number of manufacturable gates per wafer
- Relentless increase in the size of code and data storage space, encompassing solid-state, magnetic and optical storage
- Connectivity and convergence, allowing for interoperability and common functions among products from the previously separated domains of communications, computing and consumer worlds
- Expectations from customers that certain popular functions will probably spread across and become ubiquitous among product categories, such as MP3 audio and JPEG images.
- The possibilities created by Digital Video content demand an increasing flexibility of consumer appliances. The required set of applications and

their formats vary over time as standards evolve as well as per product and per country.

Consumer electronic manufacturers are expecting more from their silicon suppliers because of these trends: from flexible SoC platforms to complete solutions. This is inevitable in the age of SoCs, since so much of the system design and software/hardware partition is now done by the SoC semiconductor supplier. Therefore the industry is experiencing a re-aggregation of the design chain.

Nexperia embodies Philips Semiconductors' vision that, in order to cope with these trends, these application domains are better served by *families of flexible multimedia SoC solutions* that we call Nexperia platforms.

Nexperia platforms serve as the foundation for our customers [4] to build their novel products and services, creating the 'next experience' multimedia products that the consumers want. The Nexperia platform approach embodies the following properties:
- Flexibility (through programmability and extensive software & hardware IP choice) for easy differentiation and product upgradeability
- Innovation – addressing new, exciting, consumer applications
- Future-proof via software upgrade ability and a roadmap of compatible platform instances
- The use of a architecture framework and IP blocks to flesh out designs

2.   NEXPERIA-DVP PLATFORM CONCEPTS

2.1   Reference Architecture

Nexperia-DVP is a *Reference Architecture,* i.e. a set of documents that describes how the products of the *Digital Video Product Family* will be partitioned into subsystems and how functionality will be split over these subsystems. We normally describe Nexperia-DVP by documenting its three main parts:
a) Nexperia-DVP SoC Reference Architecture
b) Nexperia-DVP Software Reference Architecture
c) Nexperia-DVP System Reference Architecture

2.2   Standard Designs

The common subsystems, or building blocks, of the Reference Architecture are called *Standard Designs.* They can be a VLSI component (also known as device IP), a Software component, or a number of VLSI/Software components implementing a certain function.

The development of a Standard Design that implements a particular function can and is normally organized as a project, performed by a multi-disciplinary project team. VLSI and Software developers work together with Product Management to specify, develop, test and document a new standard design.

2.3   Product Platforms

When all the subsystems have been realized as Standard Designs, they can be integrated to check that the subsystems work together and implement the functions required. This set of realized, pre-integrated subsystems is called the *Product Platform*. A *Nexperia-DVP Product Platform* complies with the *Nexperia-DVP Reference Architecture*. The difference between both is that the Reference Architecture is just a document, while the Product Platform is a set of realized and integrated components.

Nexperia-DVP has been developed using the *Carrier Product* concept. Carrier Products are developed in response to fast evolving customer demands or technology opportunities and positioned in markets that are as yet fairly unknown. They are planned as first-of-a-kind products and defined as a carrier for future Product Platforms. Engineering effort is necessary to upgrade/document the product architecture to a Reference Architecture and to rework subsystems that should become reuseable assets in future projects, in order to bootstrap the Product Platform approach. Examples of Nexperia-DVP Carrier products are the PNX-8525 [5] and the PNX-8550 (a.k.a. Viper 1.0 and 2.0 respectively). This cycle is illustrated in Figure 4-1 below.

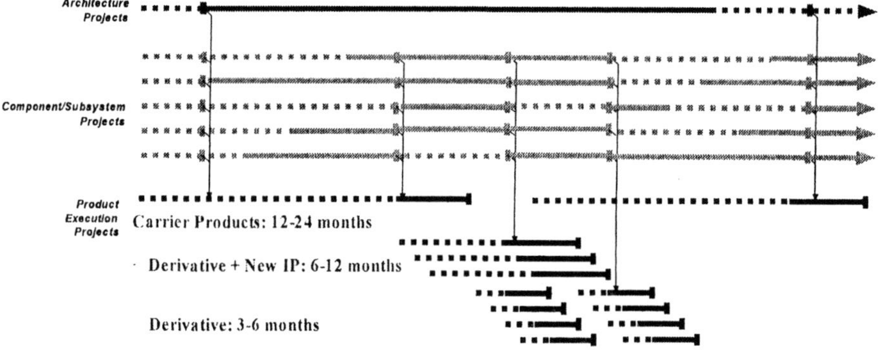

Figure 4-1. Nexperia Product Development Cycle

We expect that future developments of Nexperia-DVP will move from Carrier Products to the realization of *Product Platforms*. The new design content of a platform-based product is lowered due to the reuse of existing Standard Designs together with product-specific designs. Throughput time and risks are reduced. Platform based product realization requires that Product Platforms have been planned, created, tested and have reached a sufficient degree of maturity by the start of product realization.

2.4 Products

Nexperia-DVP Products serve on their turn as a platform to the consumer electronics manufacturers, or the foundation they use to create their own unique and novel products and services.

Nexperia-DVP products are developed rapidly and efficiently on the basis of a Product Platform, characterized by a Reference Architecture and a number of Standard Designs purchased or developed in house.

Since Nexperia-DVP products are derived from a Nexperia-DVP Product Platform, they all share properties that are valuable to our customers like flexibility through programmability, cost effectiveness, and a guarantee of being future-proof via software upgradability and a roadmap of compatible products.

3. DESIGNING NEXPERIA-DIGITAL VIDEO SOCS

3.1 Nexperia-DVP SoC Reference Architecture

Figure 4-2 illustrates the Nexperia-DVP Reference Architecture from a SoC *Connection Network* point of view. The following main elements are identified:

a) Processor Cores:

We have selected MIPS CPUs [6] to be the main control processor for Nexperia-DVP. Philips has designed and licensed a complete range of MIPS CPUs compatible with MIPS32 or MIPS64 architectures. We have also selected TriMedia [7, 8] as the main Media DSP architecture, allowing the flexible implementation of many video and audio algorithms in software. The use of MIPS as a "brancher" and TriMedia + hardware as "streamers" capitalizes on the roles that best suit each architecture (RISC vs VLSIW DSP) for a total system aggregated computational performance that can be only matched by silicon many times more expensive. Nexperia-DVP also allows for MIPS-only and TriMedia-only

Products, depending on specific product family needs for media processing in software and advanced control processing.

b) Device IP Blocks:
    The Device IP blocks implement in hardware (or micro coded HW) the necessary interfacing and processing functions for video & audio (input, output, scaling, encoding, decoding, transcoding), networking & connectivity (IEEE1394, USB, Ethernet), I/O (PCI/XIO, parallel, I2C), and architecture support (interrupt controller, semaphores, clock control).

c) Connection Network
    The connection network binds all the traffic. The diagram illustrates two main networks: the Device Control & Status Network, used to read/write to the Device IP registers, and the Pipelined Memory Access Network, used by the CPUs and Device IPs to access the main memory.

d) Main Memory Interface (MMI):
    Connects the Nexperia-DVP SoC to the main memory, typically a high-speed DRAM memory.

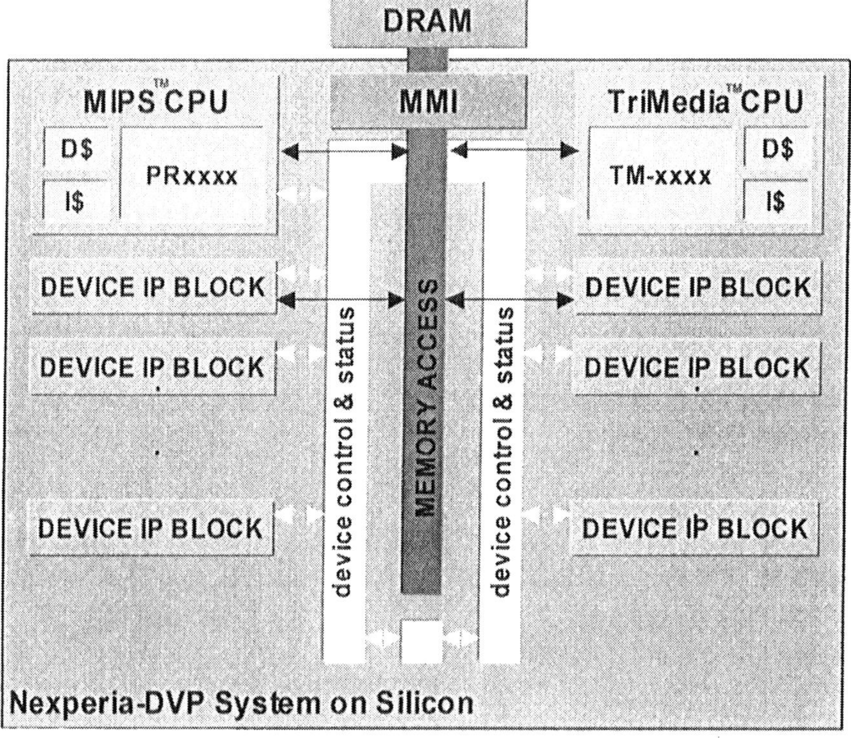

Figure 4-2. Nexperia-DVP Reference Architecture

3.2 Three Levels of Abstraction

We have defined the Nexperia-DVP SoC Reference Architecture using three levels of abstraction. These levels were used in order to avoid traditional architecture definitions that do not cope with evolution because the Device IP library is tied to a particular bus selection. The three Levels of Abstraction are:
- Level 1: Software-Hardware Platform Rules
- Level 2: Device Transaction Level
- Level 3: Connection Network

3.2.1 Level 1: DVP Software-Hardware Rules

This level deals with the software view of the hardware. The "DVP Software-Hardware Rules" defines:
- Unified Memory Architecture
- Rules for Data Movement
- Endianess
- Ordering & Coherency
- Interrupts
- Data formats, including Pixel Formats.
- Trimedia-MIPS Communication
- Protection
- Boot

The Unified Memory Architecture specifies that all addressable objects in the system should be identified by the same address. We have defined the two main types of data movement: Device Control & Status (DCS) transactions that access 32-bit only registers in the Device IP and must have no side effects on read, and Direct Memory Access (DMA) transactions that move data to/from memory to the Device IP.

3.2.2 Level 2: Device Transaction Level (DTL)

Level 2 deals with point-to-point transfers between Device IPs and the Connection Network, specifying a Device IP *partition* and *architecture* that is compatible with Level 1. Two main types of ports are typical of Device IP:
- Device Control & Status Ports: these are read/write to memory mapped control and status registers in the Device IP
- DMA ports: the Device IP uses these ports to communicate to memory. It also allows for direct Device-IP-to-IP communication.

With the introduction of DTL ports and protocol to our Device IP library, we managed to:

−   Allow the reuse of the Device IP even when the Connection Network evolves.
−   Remove details of the bus protocol from IP development, letting the IP move data in the most natural way for that IP
−   Move any clock domain boundary out of IP and into the bus interface

The Nexperia-DVP concept of DTL ports and DTL protocols precedes and resembles VSIA's VCI interface [9], but it has an important conceptual difference: While the VCI interface was meant to just abstract the specific type of bus utilized in the systems from the IP, with DTL we abstract the system aspects (memory bandwidth, optimal transaction lengths for the memory and the systems, amount of buffering needed, etc.) from the Device IP. Figure 4-3 shows an example of our Device IP architecture partition.

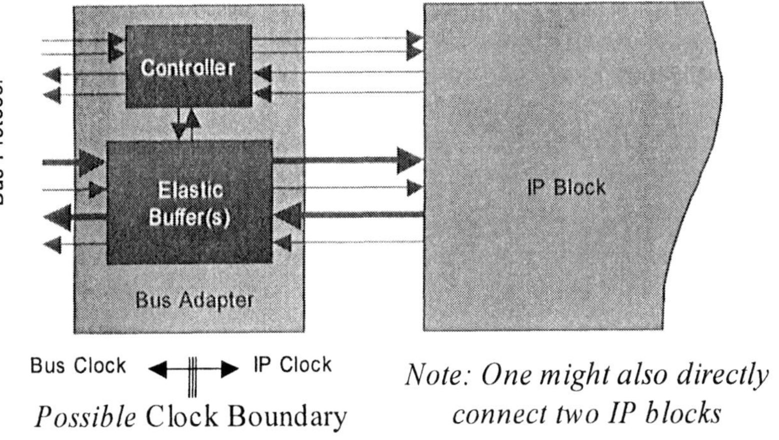

Figure 4-3. Device IP Architecture Partition

3.2.3   Level 3: Connection Network

The Nexperia-DVP Level 3 Connection Network deals with the more traditional Bus Hierarchy & Bus Level: it specifies a bus hierarchy and for each bus their wires, clock cycle protocols, AC characteristics, hardware architecture & hardware blocks.

The second generation Nexperia-DVP is composed of:

−   The Device Control & Status (DCS) Network
−   The Memory Connection Network

3.2.3.1 Device Control & Status (DCS) Network

From a system/SoC architecture point of view, the DCS Network is primarily a low latency communication path for the CPUs and other initiators to access the control & status register in the Device IPs. From a VLSI physical architecture point of view, the DCS Network allows for the implementation of a Physical Design Strategy of *"islands of synchronicity"* that we will describe later in this section. A DCS Network has the following properties:

- 8, 16 and 32 bit transactions
- Time-out generation and multiple bus system design
- Sampling of bus signals in case of error or time-out generation
- Both posted and precise writes
- Low power design
- Protection with selective blocking of initiators access to targets
- Signals and protocol very similar to DTL, allowing for simple and efficient adapter design
- Compatible with "chiplet" (or island) physical design approach
  - Synchronous and asynchronous interconnect options
  - Efficient layout with few top level wires and minimum netlist partitioning requirements
  - Fast timing closure

Figure 4-4 shows the logical view of a DCS network. Typically each Device IP has a DTL-based DCS port that is hooked to the DCS Network via a DTL-DCS adapter. Device IPs are designed such that DTL DCS ports can operate at the IP Clock (from 70 to 150 MHz) or at the DCS Network clock (currently at 200 MHz).

A DCS Network Controller is configured to each initiator/target characteristics, such as the clocking mechanism (synchronous or asynchronous), and the type of register write allowed (posted or non-posted writes). The DCSN Controller implements round-robin arbitration between initiators, the splitting of command to targets (address decode), the collection and multiplexing of response from targets, the timeout generation, error handling (captures error transaction), and protection.

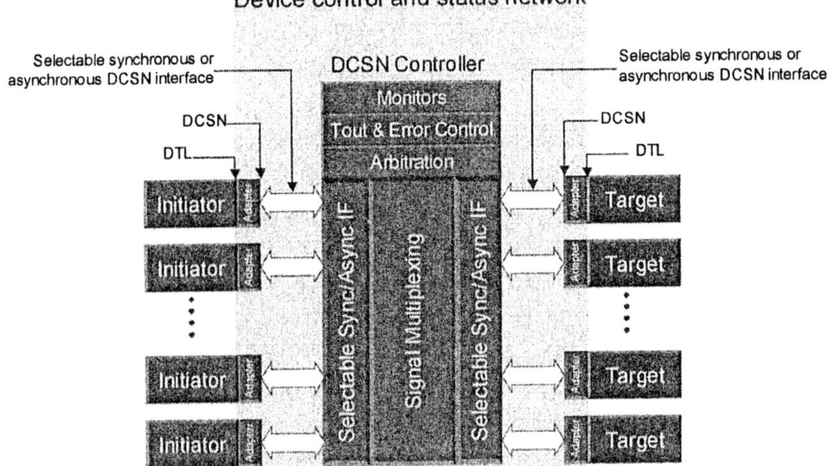

Figure 4-4. Device Control and Status Network

3.2.3.2   Memory Connection Network

Nexperia-DVP SoCs Memory Connection Networks are expected to evolve over time, and we will automate the generation of such structures. The current generation is implemented with two key elements: Memory Transaction Level (MTL) ports and protocol and the Pipelined Memory Access Network, that are both described below.

MTL is a communication protocol optimized for communication to a (DRAM) memory controller. Figure 4-5 shows the application of MTL ports in a typical Nexperia-DVP SoC. CPUs as well as DMA agents communicate to memory using MTL. MTL is a point-to-point interface protocol, which means that CPUs typically have their own private MTL connection to the memory controller. In the DMA memory infrastructure MTL is used to connect DMA adapters to the connection network and to interface the connection network to the memory controller. In mid-range/low-cost applications there might be one DMA adapter that interfaces directly with MTL to the controller. DTL-MTL Adapters are required to translate Device IP style transaction to system/memory optimized transactions.

We have implemented the current generation of our Memory Connection Network via PMAN (Pipelined Memory Access Network). From a system/SoC architect point of view, the PMAN is primarily a high bandwidth hierarchical communication path for the Device IPs to communicate to memory. From a VLSI physical architect point of view, the PMAN allows for the implementation of a Physical Design Strategy of "islands of synchronicity" and hierarchical split that we will describe later in this session.

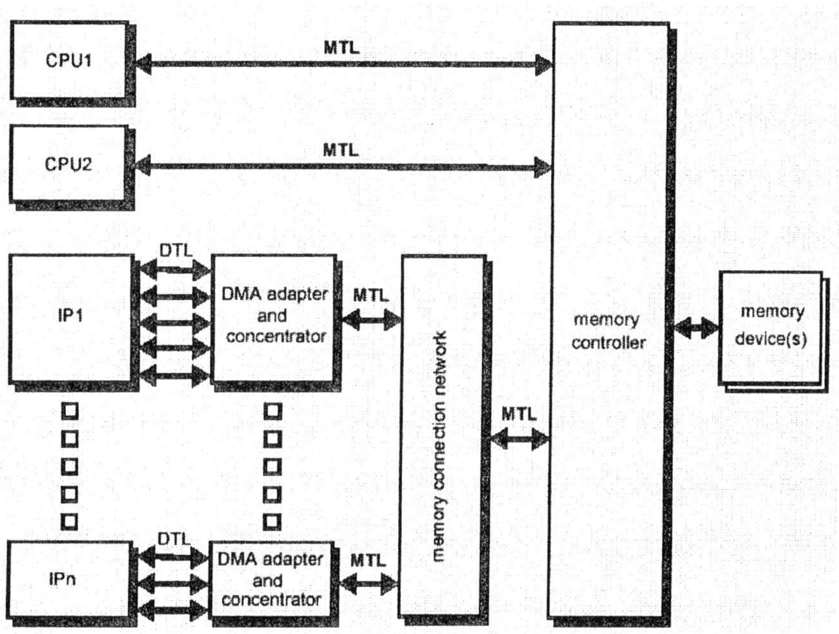

Figure 4-5. Application of MTL Ports in a typical Nexperia SoC

PMAN allows the implementation of a mixture of deferred/local arbitration; non-symmetrical hierarchies for write and control data muxing; MTL to MTL routing; memory protection via the partition of memory into "sand boxes". Each MTL initiator is allowed or denied access to each sand box. Current PMAN implementation is clocked at 250 MHz and allows a maximum data rate of 1 Gbytes/second.

3.3 SoC Implementation

As we have seen, the Nexperia-DVP SoC Architecture strives for the orthogonalization of the Communication Architecture versus Device IP Functional considerations. This approach and the characteristics of the protocols selected are with respect to the SoC implementation:
– Physical Hierarchy Friendly
– Timing Closure Friendly
– Handoff Friendly
– Change Friendly
This allowed for a SoC Physical Implementation that is characterized by:

a) Physical hierarchical partition in "chiplets"
b) Islands of synchronicity

Partitioning divides the logic hierarchy into manageable sized blocks, called chiplets. Modules (processors, IP devices and the Communication networks) are divided over the chiplets to create Islands of Synchronicity within the chiplets. Inter-chiplet communication is either asynchronous (DCS), or flop-to-flop (PMAN).

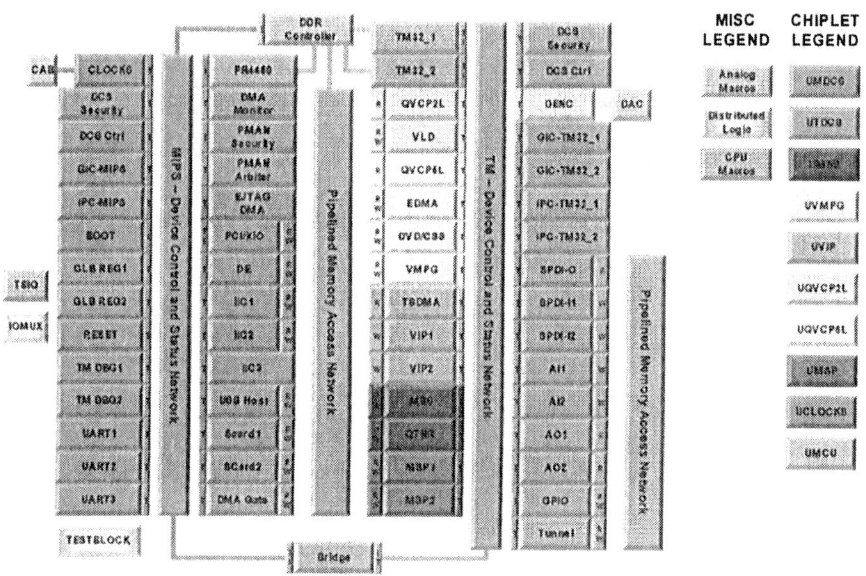

Figure 4-6. Logical to Chiplet Mapping

Closely related to top-level floorplanning, partitioning follows these guidelines:
- The logical hierarchy, characterized by the DCS and PMAN network dataflow in the design
- Size and complexity of modules
- Clock domains within modules; clock domains (except PMAN clock) are within one chiplet, making it locally synchronous.
- Chip IO Timing requirements of the modules

Signals connect between the chiplets via abutment. There is no top-level routing, except for the clocks that had to be matched. The inter-chiplet signals between non-neighboring chiplets are routed by inserting feed-through buffers in the chiplets that the signal had to route through. Figures 4-6 and 4-7 show the logical to chiplet mapping and the resulting floorplan for the PNX-8550. A more detailed description of the implementation of our first generation of a Nexperia-DVP product is described in [10] and [11].

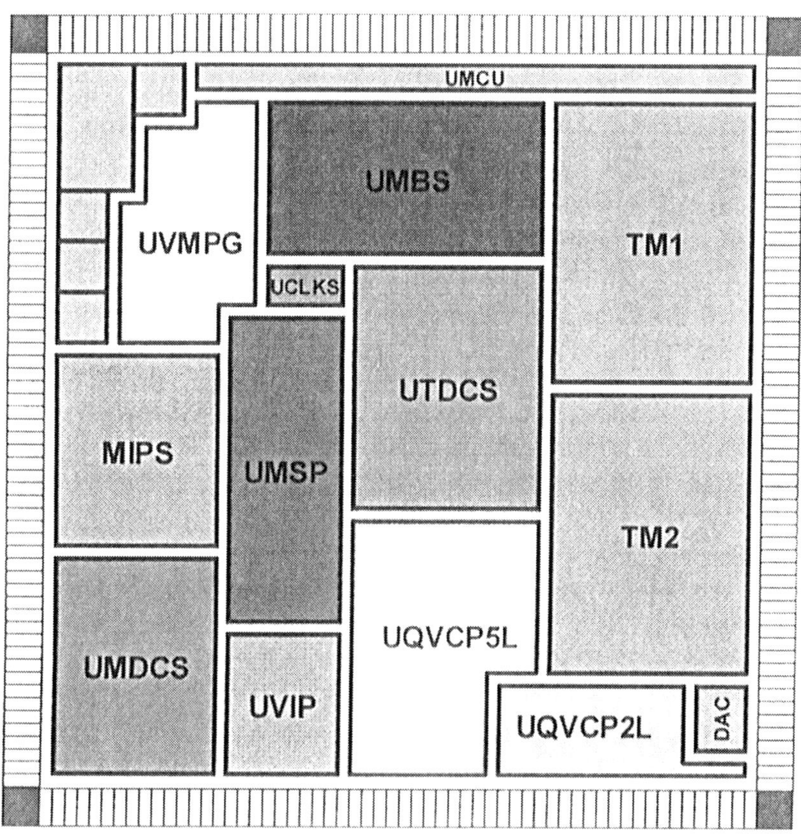

Figure 4-7. Floorplan for the PNX-8550

4. NEXPERIA-DVP SOFTWARE ARCHITECTURE

In this section we will cover a brief overview of the Nexperia-DVP software architecture. A thorough discussion of the entire software architecture for Nexperia-DVP products is beyond the scope of this chapter, so we will introduce only those concepts and parts that are most relevant to hardware/software co-design challenges that will be covered in the next section.

4.1 Nexperia-DVP Platform Software

At the top-level, a Nexperia-DVP product typically consists of three large subsystems: the operating system, the Nexperia-DVP platform software and the customer's middleware and applications.

- *Operating System:* In principle the Nexperia-DVP architecture can function with any operating system on its main control processor. The operating systems supported by Philips Semiconductors, however, are WindRiver's VxWorks and Linux. It is the explicit strategy of Nexperia-DVP to make the best and widest possible use of the services and libraries offered by the operating system's vendor. This not only includes real-time-kernel services (i.e. task scheduling) but also graphics, user interface management systems, protocol stacks, file systems etc.
- *Nexperia-DVP Platform Software*: The Nexperia-DVP platform software, developed and maintained by Philips Semiconductors basically deals with all handling of audio and video streams. It spans multiple processors and covers codecs and other DSP routines as well as sophisticated filter graph management, synchronization and buffer management. The Nexperia-DVP platform software offers an extendable, standardized programming interface (API) for most functions, and is programmed using standard industry practices [12]. The Nexperia Development Kit (NDK) allows customers and partners to develop software for the platform and is provided with the platform software.
- *Customer's Middleware and Applications*: As the name implies, Philips Semiconductors generally does not offer more than platform software and operating system as described above (although counterexamples do exist for some markets). The middleware running on Nexperia-DVP products spans a huge variety of programming languages, vendors, sizes and application domains, so further generic discussion is not possible.

4.2 Multiple Processor Software Consequences

The Nexperia-DVP platform software often spans multiple processors in a single system. These may be general purpose CPUs (either on-die or off-die), or specialized VLIW audio/video DSP CPUs (TriMedia's [5]). All these CPUs utilized are plain-C programmable, and the Nexperia-DVP software architecture has a high degree of transparency and uniformity across processors.

Most inter-process communication is handled via remote-procedure-calling (RPC), with automatic proxy/stub generation. Because most components in the Nexperia-DVP stack are available for execution on any processor, this provides almost complete functional transparency with respect to which component runs on which CPU.

Of course, the execution and performance impact of partitioning tasks over CPUs is still very large, and the systems contain CPUs specialized towards particular tasks. So any Nexperia-DVP product typically contains a fixed partitioning of tasks over CPUs that is carefully designed up front. Dynamic balancing of loads and transferring of tasks is implemented only in some products, and then in a hard-coded, pre-designed fashion.

4.3   Streaming Component Architecture

Nexperia-DVP deploys a two-level component architecture for its streaming software: standard components and functional subsystems.

Standard Components: All audio and video streaming components in Nexperia-DVP follow a set number of coding and functional standards, collectively known as the Trimedia Software Streaming Architecture (TSSA). Of course the TSSA standard provides for different classes and types of components such as implementations of codecs, hardware accelerator drivers, buffer managers, and filter graphs. The TSSA architecture is internally layered; this is not covered further in this text. Examples for TSSA components are an MPEG Transport Stream (TS) demultiplexer, an MPEG2 video decoder and an AC3 audio decoder.

– ***Functional Subsystems:*** From this library of standard components a number of larger functional subsystems are built that can be combined into a product. These subsystems are generally aligned with the top-level product functions and features as requested by customers. Examples are a DVD playback subsystem; an ATSC (Advanced Television Standards Committee) broadcast decoding subsystem, etc. It is important to note that different functional subsystems can contain (different instances of) the same component, e.g. both the ATSC player and the DVD player mentioned above use the same MPEG2 video decoder.

4.3.1   Reference Architecture

A reference architecture is deployed to describe the different possible combinations of functional subsystems as described above. This reference architecture is illustrated in Figure 4-8. The three main classes of functional subsystems in this diagram are players that decode compressed audio/visual (A/V) streams, recorders that encode uncompressed A/V streams and transformers that translate compressed A/V streams into different compressed A/V streams.

The Nexperia-DVP software reference architecture has an uncompressed A/V part (top-most part of Figure 4-8), which includes analog A/V input subsystems and a presentation engine containing all audio and video processing like sound equalizing, improvement and mixing, and video improvement, scaling and pixel/frame rate conversions. Since performance requirements are most stringent in especially the uncompressed video domain, Nexperia-DVP utilizes several hardware acceleration options for these operations.

The uncompressed A/V parts of the reference architecture (bottom 2/3rds of Figure 4-8), consists of a flexible network of five classes of functional subsystems:

- Uncompressed (Digital) Inputs: The subsystems contain the driver for a digital interface, either broadcast (push-mode) or networked or storage (pull-mode), and potentially any associated protocol stack s/w. Digital inputs deliver compressed A/V streams over a unified interface to any player (depending on format of course).

- A/V Players: These subsystems contain relevant components for stream demultiplexing, audio and video decoding, synchronization and any system information processing as required. Players can take their input from any uncompressed input, or from any recorder/transcoder through a loopback.

- A/V Recorders: These subsystems take uncompressed audio and/or video and encode this into a compressed A/V stream. They contain relevant components for encoding and stream multiplexing.

- Transcoders: These subsystems take an uncompressed A/V stream in, and generate another uncompressed stream out, which is in a different format, has a selected service, a different bit-rate etc. Transcoders can be implemented as full decoder/encoder combinations, or as dedicated transcoding implementations.

- Uncompressed (Digital) Outputs: These subsystems are the counterparts of the uncompressed inputs, and contain driver and protocol stack for the interface in question. For read/write capable interfaces (e.g. HDD or Ethernet), the uncompressed input and output subsystem may actually be implemented as a single subsystem.

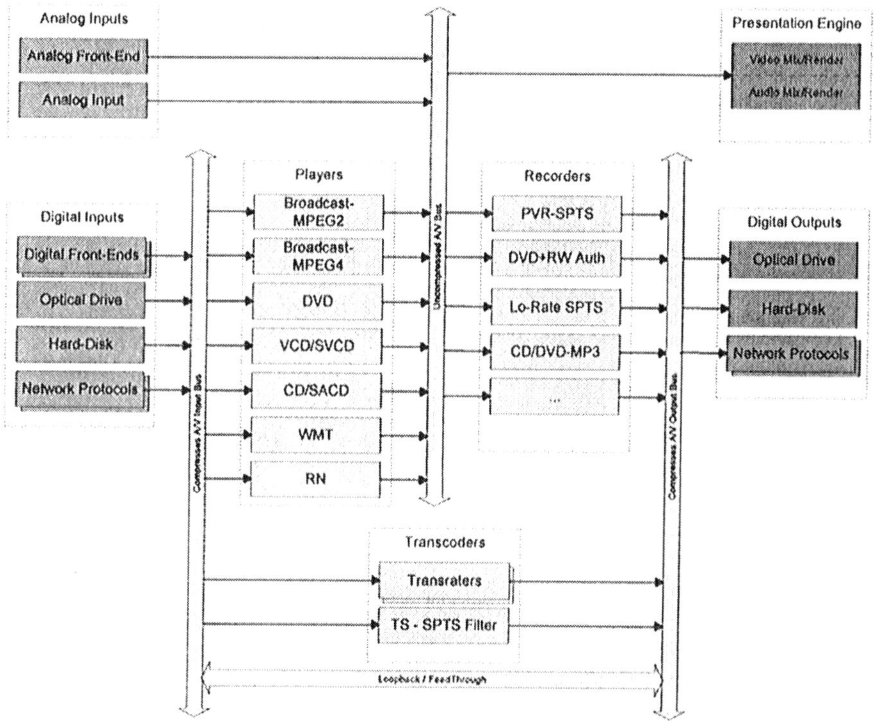

Figure 4-8. Nexperia-DVP Software Reference Architecture

5. NEXPERIA-DVP SYSTEM INTEGRATION

In the previous section we discussed techniques deployed in Nexperia-DVP to manage the large diversity of product types and instances, product family members and product use-cases that occur in the Nexperia-DVP family. In this section we will discuss how to make this variety of systems and operating modes actually work and achieve optimal performance characteristics.

As is the case with all embedded digital software architectures, getting it right with respect to the execution architecture of the entire Nexperia-DVP platform has proven to be one of the most challenging achievements in the Nexperia-DVP software architecture. Nexperia-DVP deploys a collection of sophisticated mechanisms that together allow a near optimal utilization of system resources, retaining excellent product stability and robustness.

When reading this chapter, one must realize a basic 'rule of thumb' that holds for Nexperia-DVP systems, as well as many other systems: resource overkill can make the execution architecture challenge an almost trivial one.

Roughly speaking, the basic architecture mechanisms provided in TSSA for Nexperia-DVP products require little/no additional 'tuning' if usage of critical resources (memory, time etc) is not required to exceed an average of 60-70%. Only when one or more available resources must be utilized to their fullest potential (even if only in some use cases) will it become significantly more complex to make each product and each use-case work.

The consumer electronics markets, where many Nexperia-DVP products are used, often require highly maximized usage of all available resources, which makes execution architecture a challenging and important topic for Nexperia-DVP. This is a direct consequence of balancing the cost limitations imposed by a mass consumer market and at the same time pushing the performance envelope with advance applications.

The following sections will elaborate on the basic principles of Nexperia-DVP execution architecture, and provide some quantitative guidance on the basic parameters.

5.1   Performance Characteristics

The following are the four 'core' performance requirements in most Nexperia-DVP systems:
- ***The system must work under normal conditions:*** The obvious one, although often the hardest to achieve.
- ***The system must have a given (short) maximum input/output delay:*** A long input-to-output delay not only translates into a lot of memory usage, it is also unacceptable for many audio/video consumer products for reasons of multiple product synchronization issues, user feedback and reaction time issues etc. Many Nexperia-DVP video products require (and achieve) input to output delays of one field or less, where a field refers to one-half of the TV frame that is composed of either all odd or even lines. In CCIR systems each field is composed of 625/2 = 312.5 lines (20ms), in EIA systems 525/2 = 262.5 lines (16.66ms). There are 50 fields/second in CCIR/PAL (European standard), and 60 in the EIA/NTSC (North American) TV system.
- ***The system must behave gracefully under exceptional conditions:*** Under conditions of erroneous inputs, inputs exceeding bandwidth specifications, misbehaving other parts of the system, etc., the Nexperia-DVP platform must show predictable and acceptable behavior. Not only should the system not crash, it should not display unacceptable artifacts on audio or video outputs.

– ***The system must be as cost effective as possible:*** This requirement is very difficult to quantify, and can only be measured against competitor offerings in the market. However, the high price sensitivity and high product volumes of the markets in which Philips operates with Nexperia-DVP make spending effort to optimize and maximize system performance a reality. Of course this requirement remains a trade-off between development effort and unit cost (of Nexperia-DVP chip and additional external components).

5.2 Critical Resources

In Nexperia-DVP products the following three resources are regarded as the most 'scarce' or 'expensive' and are managed accordingly.

– ***Memory Bandwidth:*** This is the most critical quantity for many Nexperia-DVP products. Due to its unified memory architecture, all general-purpose CPUs and hardware accelerators share access to a single memory bus connected to external DDR RAMs. Ability to maximize memory utilization translates directly into cheaper memories. Current Nexperia-DVP systems typically can use 16-bit 133MHz DDR RAMs providing 533MB/s raw bandwidth, up to 32-bit 225MHz DDR RAMs providing 1.8GB/s raw bandwidth or more.

– ***CPU Cycles:*** The second most critical quantities for most systems are the CPU cycle budgets of the various CPUs in the system. This poses a 'classical' process scheduling challenge, as well as a complex budgeting challenge due to the interaction between memory bandwidth and CPU performance (further explained below).

– ***(RAM) Memory Size:*** The third and final of this list, memory size obviously translates directly to product cost (given the use of external memory). Due to the streaming nature of most Nexperia-DVP products, memory consumption for video buffers often accounts for half or more of the total system memory usage. In particular, the number and size of audio/video buffers required can be improved by optimizing the system. Typical memory sizes currently are 32 and 64 Megabytes.

In addition to the three quantities mentioned above, Nexperia-DVP deploys techniques to budget, schedule and manage, amongst other factors, network channel bandwidth, flash memory size and power consumption. These topics are beyond the scope of this text.

5.3 Bandwidth and CPU performance

One of the complicating factors in designing optimized schedules for unified memory products like Nexperia-DVP are the interaction between memory bandwidth consumption and CPU performance. Primarily this is due to the effects of hardware accelerator and 'other' CPU accesses to memory on average memory transaction latency. Let's take a closer look at the requirements CPUs and hardware accelerators have on memory bandwidth and latency.

- ***Hardware Accelerators:*** Most accelerators used in Nexperia-DVP are asynchronous in nature and have an internal clock frequency significantly higher than that required for the systems in which they are used. All accelerators, even the synchronous ones, deploy FIFO buffers in their DMA channels, resulting in a rather low sensitivity to memory transaction latency.

  As long as sufficient bandwidth can be made available during the desired periods of execution the blocks will meet their performance requirements. Obviously, as the desired period of execution approximates that minimum execution time (determined by the accelerators internal clock frequency), the sensitivity of the block to memory starvation increases.

- ***CPUs:*** The CPUs deployed in Nexperia-DVP products (MIPS and TriMedia) all include both instruction and data caches. Nevertheless any CPU remains highly sensitive not only to available bandwidth, but especially to the memory transaction latencies experienced when consuming its bandwidth allocation. In current Nexperia-DVP products with CPUs in the 300MHz range and 32-bit DDR memory in the 250MHz range, transaction latencies for a 128-byte cache-line refill measure from 30 to 100+ CPU clock cycles. The increasing divergence between CPU and memory speeds will only compound this in the future.

5.4   Effect of bandwidth on transaction latency

The memory transaction latency experienced by a CPU is influenced by four different factors:

- ***Minimum transaction cost:*** The minimum length of a CPU memory transaction is composed of three factors: cycles required in the bus and arbitration system, actual memory access and DDR access penalty cycles. These factors together give a minimum average transaction length that is achieved only if no other blocks access the memory. For most Nexperia-DVP products this is in the 20-40 CPU cycles range.

- ***Transactions of other blocks that take precedence:*** Some hardware accelerators, or other CPUs will have the ability to issue transactions that take precedence over the CPU's pending requests. This can be due to slotting, priority or other effects (see DDR arbitration below). These transactions are the dominating factor in determining the additional memory access penalty.

- ***Pending transactions of blocks that do not take precedence:*** Transactions that would not take precedence over the CPU when arriving *are* completed when pending at the time of CPU transaction request arrival. In Nexperia-DVP systems it is possible to interrupt such transactions, but this option is seldom used due to the bandwidth loss it causes (generated by higher DDR inefficiency).
- ***Other transactions of the CPU itself:*** Higher bandwidth consumption of the CPU itself does not by itself increase its memory transaction latency. However, interaction between the buffering DMA patterns of hardware accelerators and high CPU transaction volumes does move hardware accelerator transactions out of the 'quiet zone': the period right after a CPU memory access when it is using its newly acquired cache line(s).

The intricate nature of the interaction between these factors makes simple stochastic models highly inaccurate, especially in the corners of the performance envelope. Simulations of bus behavior under realistic conditions show that transaction latency increases dramatically when (precedence taking) DMA load exceeds 60-70% of the available bandwidth. This is shown in Figure 4-9.

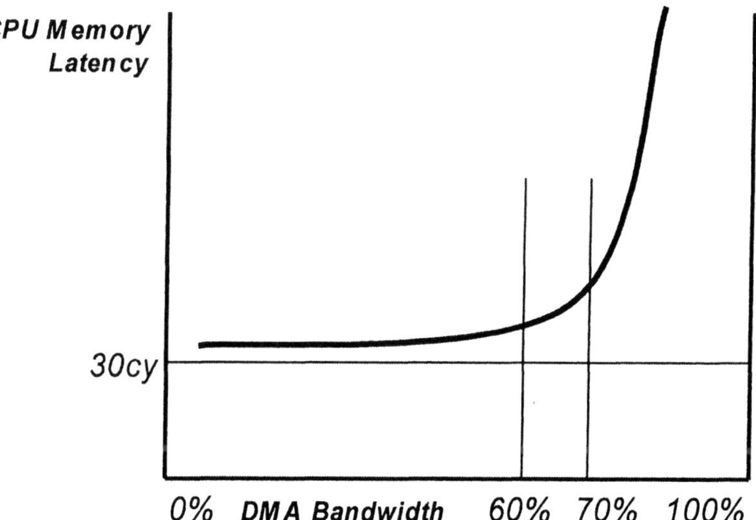

Figure 4-9. Memory Latency vs. DMA Bandwidth

5.5   Effect of transaction latency on CPU performance

The effect of long memory transaction latencies on CPU performance is governed primarily by a CPUs cache miss rate. The type of CPU, type of software and size of caches used influences this miss rate significantly.

In Nexperia-DVP code is classified as either control or DSP-type. This classification can be done at various levels of granularity (task, file, function, statement) depending on the accuracy required.

- *DSP-type code:* This type of code generally manipulates actual audio and/or video samples. Examples are MPEG codecs, filters, scaling operations etc. Roughly speaking this code is characterized by long complex expressions and frequently repeating tight loops. Also the average working set of a DSP task is small, often much smaller than the CPUs I-cache size.

- *Control-type code:* This type of code generally deals with controlling the hardware and software (DSP) operations on audio/video or with other parts of the system. Roughly speaking this code is characterized by lots of decision-making (switch/if-statements) and function calls. The average working set of a control task (or control portion of a task) is often very large, often much larger than the CPUs I-cache size.

Table 4-1. Characteristics of DSP and Control Code

	DSP code	Control code
Characteristics	Long expressions Highly repetitive loops	Lots of switch/if statements Many function calls
Working Set Size	Typically smaller than CPUs I-cache size	Typically much larger than CPUs I-cache size
Typical instruction repetition rate before cache line invalidate	40 or more	Between 1 and 3
Typical CPU stall cycles due to cache misses on moderately loaded system	20% and below on TM	80% and above on TM 60% and above on MIPS
Predominant Cache Misses	D-Cache	I-Cache
Typical bandwidth consumed for each <u>effective</u> CPU Mcycle (i.e. excluding cache-stall cycles)	400KB/s on TM	6.4MB/s on TM 1.5MB/s on MIPS

Typical behavior of control and DSP tasks differs dramatically on both general purpose and VLIW CPUs. The preceding table (Table 4-1) illustrates the typical average behavior of these types of code on a Nexperia-DVP system. Of course, individual tasks and systems vary widely in their expressed behavior and are generally measured on a task-by-task basis.

From the numbers in the table above one can see that increasing memory transaction latency has a dramatic effect on the performance of the CPUs, especially in the case of control code execution. The DMA load at a specific point in time may influence the time required to complete a control task by a factor of two or more!

5.6 Effect of CPU performance on system scheduling

As all blocks in a Nexperia-DVP system except input and output blocks are controlled and triggered by software, the performance of the controlling CPU on these control tasks influences the timing and schedule of execution of these blocks. This brings us full circle: the scheduling of on-chip, hardware-based function accelerators influences memory bandwidth; memory bandwidth influences memory transaction latency; memory transaction latency influences CPU performance; and finally, CPU performance influences scheduling.

As said before, only when achieving bandwidth and CPU utilizations exceeding 60-70% do these interactions have to be completely modeled and measured. This is the case for many Nexperia-DVP products and the markets they serve.

The following sections will discuss the basic memory arbitration and scheduling techniques deployed, and then show the process used in Nexperia-DVP to obtain a working schedule for a use-case.

5.7 Memory Arbitration

Many use-cases in Nexperia-DVP products have, either probabilistically or deterministically, periods of execution in which the total amount of bandwidth requested by all CPUs and hardware accelerators exceeds the available bandwidth. To regulate the distribution of available bandwidth over the requesting blocks Nexperia-DVP utilizes an advanced DDR arbitration scheme that aims to achieve two goals:
- Fair distribution of bandwidth over requesting blocks
- Shortest possible transaction latency for CPUs

From a software point-of-view, the techniques deployed to achieve short CPU transaction latency are, although very important, of little interest in the execution architecture design process, since they require little or no tuning.

Most noteworthy is the decision whether or not to allow interruption of pending memory transactions to shorten CPU transaction latency. As said before, in most systems today this option is not used due to the additional bandwidth loss it causes.

The Nexperia-DVP memory arbiter deploys a sophisticated algorithm for distribution of bandwidth over requesting blocks. The essence of what is offered for software control, however, is simple, and constitutes two principles:

- ***Guaranteed bandwidth:*** In Nexperia-DVP it is possible to guarantee a specified amount of bandwidth to each individual block (CPU or hardware accelerator). This amount can be less than, equal to, or more than the amount needed/requested by the block at any point in time. Bandwidth guaranteed but not used by any block is added to the remainder pool, which is distributed on priority basis (next paragraph). If a block requests more bandwidth than has been guaranteed, the block will either be stalled (i.e. run slower), or consume additional bandwidth from the remainder pool.

  Bandwidth guarantees are necessary primarily to guarantee proper operation of synchronous hardware accelerators (mostly input and output blocks), and accelerators that must run close to their maximum performance level (i.e. cannot handle a lot of stalls). Because of this Nexperia-DVP implements bandwidth guarantees within a fairly short period (several 100 cycles). This, in turn, strongly influences the CPU transaction latency. As a general rule of thumb, bandwidth guarantees are kept as low as possible to ensure proper operation of the system.

- ***Priority based distribution of remainder:*** In Nexperia-DVP the total sum of all bandwidth guaranteed to blocks is always significantly less (20%-70%) of the total available. The remaining bandwidth that has not been guaranteed plus the guaranteed bandwidth that has not been consumed is distributed over CPUs and accelerators in a priority order. Blocks may appear both in the guarantees list and the priority list, and frequently do.

5.8   Scheduling Techniques

In Nexperia-DVP several different scheduling techniques [13] are used to generate working system. We will first discuss the basic task scheduling mechanisms used, then we will study the scheduling of hardware accelerators, and finally we will discuss the actual approach used in current Nexperia-DVP products.

5.8.1   Scheduling of Tasks

Although Nexperia-DVP deploys standard embedded operating systems (pSoS, VxWorks, etc), it uses multiple additional mechanisms to schedule the various tasks on the CPUs.

- ***Priority based scheduling:*** This is the standard mechanism offered by the operating systems used. The various tasks in the system are all assigned a priority, either statically or dynamically, and execution is in priority order. Standard techniques like Rate Monotonic Analysis are used to determine working schedules.
- ***Reservation based scheduling:*** This technique assigns a CPU cycle 'budget' to a task or a set of tasks. Tasks are scheduled so that each tasks gets its assigned budget. This technique is combined with priority based scheduling by first assigning a budget to a group of tasks and then using priority based scheduling within this group.
  Usage of this technique is under development in Nexperia-DVP. It is expected to be most useful in achieving a higher level of robustness in use-cases dealing with multiple media streams (where each stream would be individually budgeted).
- ***Hard coded scheduling:*** If a group of tasks is required to achieve extremely tight schedules (with respect to memory size, bandwidth or CPU performance), it is sometimes necessary to completely plan out a schedule and execute it in a hard-coded fashion, i.e. in a fixed task-to-task order.
  In Nexperia-DVP this technique is used in the uncompressed video domain, where typically the number of operations is very small (less than 10), the consumption rates are very high (several 10%'s), and the requirements tight.
  This technique is significantly more robust and reliable than the other two, but also less flexible. The Nexperia-DVP methodology contains methods and tools to ease the complexity of designing and implementing hard-coded schedules.

5.8.2   Scheduling of Hardware Accelerators

Although scheduling of tasks as discussed above is important *per se*, it is important to note that a typical audio/video flow graph in a Nexperia-DVP system contains a mix of software (DSP) operations and hardware-accelerated operations. Almost all hardware accelerators in Nexperia-DVP can process up to a complete field/frame without software intervention. This makes determining a working schedule in essence a many-processor scheduling problem.

Some special techniques are used to make use of this parallelism potential (see also Figure 4-10):

- ***Sequencing of operations:*** This is the simplest of techniques. Essentially when one operation is executing on frame/field number N, the previous operation starts executing on frame/field number N-1. In this way two processors never operate on the same frame/field (in the temporal sense) at the same time.
- ***Slicing:*** Many hardware accelerators and DSP routines in Nexperia-DVP support operating on parts of a frame/field. This operating mode is called slice based processing. When two processors are concatenated and are both running slice-based, scheduling can be based on slices too. Basically, if processor A has completed its first slice of frame N, it is handed over to processor B. Now processor A starts executing on slice 2 at the same time that processor B starts executing on slice 1 of the same frame/field.
- ***Staggering:*** When two processors A and B are concatenated and are both sequential and localized, i.e. they read their input and write their output front-to-back, it is possible to start processor B before processor A is finished with the same frame. If the delay in starting processor B plus the minimum execution time of processor B is at least the maximum execution time of processor A, it is guaranteed that processor B will never overtake processor A on the same frame. Obviously, successful use of this technique depends on the ability to accurately predict minimum and maximum execution times of all processors in the system.

Obviously the techniques above require very different software control. Slicing is significantly more expensive than sequencing and staggering due to the higher task-switch or interrupt frequency it causes (software intervention is required to trigger every slice of both processor A and B). If the slices are small enough this becomes very significant. Staggering is not much different from sequencing in amount of software control effort required, but requires accurate time/threshold delays as a concept in the architecture.

It is also worth noting that the different techniques above have a different influence on the amount of memory required for audio/video buffers. When using sequencing, two full frame/field buffers are required between each pair of operations. When slicing still two buffers are required between each pair but they may be one slice in size. When staggering, most pairs can do with only one full frame/field buffer. Of course, actual buffers required must be studied on a case-by-case basis, and depends on the algorithms implemented by the processors and the history they require.

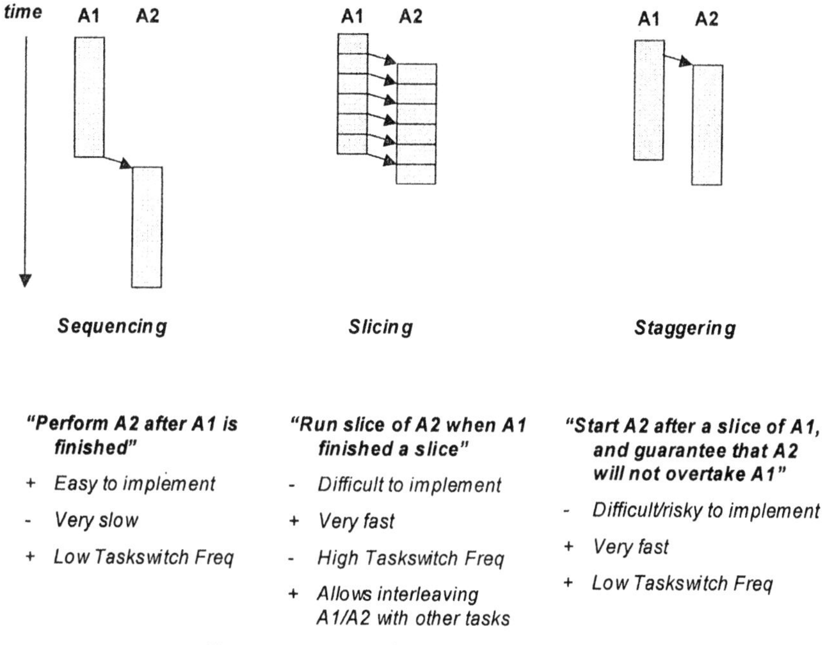

Figure 4-10. Special Scheduling Techniques

The following table summarizes the properties of these techniques:

Table 4-2. Properties of Scheduling Techniques

	Sequencing	**Slicing**	**Staggering**
Software effort	Lowest	Higher, depends on slice size	Low
Buffer memory	Largest	Smallest	Middle
Input/output delay	Longest	Shorter	Shorter
Requires of processors	Nothing	Must support slicing	Must be sequential and localized

5.9   Putting it all together

The approach used in Nexperia-DVP can be summarized as follows:
- The uncompressed video domain, i.e. all operations on uncompressed video buffers are singled out as 'big ticket items'. In most Nexperia-DVP systems the uncompressed video operations account for >60% of

memory size, bandwidth and CPU budgets.

For these operations the following techniques are used:

    a) A hard-coded schedule is designed that correctly sequences all operations with an acceptable bandwidth profile.

    b) Operations are sliced to reduce memory consumption, usually half or quarter field/frame.

    c) Slice operations are then staggered to further shorten input/output delay and memory consumption.

– The remainder of tasks runs in the remainder of CPU cycles and memory bandwidth, using the following techniques:

    d) Basic priority based scheduling. Reservation based scheduling is being developed as an additional technique.

    e) Simple sequencing of operations.

Now we have discussed all techniques used in Nexperia-DVP to develop working schedules for specific use-cases, we can summarize the process of developing such schedules. We will describe the process for a single use-case. Additional implementation techniques present in Nexperia-DVP to manage the diversity of, parameterization of and switching between use-cases are beyond the scope of this text. The process consists of two iterative steps.

5.9.1   Step 1: to determine the uncompressed video schedule and bandwidth profile.

First the flow graph of all uncompressed video operations is determined, and raw data for each operation measured. This includes clock cycles required, bandwidth required and, for CPUs, cache miss profiles and counts.

Then a schedule is designed, by hand, that sequences in time the various uncompressed video operations using sequencing, slicing and staggering as described above.

Tools are available to determine the resulting timing and bandwidth profiles of the schedule design. Usually the system either does not meet its deadline or exceeds bandwidth requirements, or both, and the process becomes an iterative manual process of tuning the schedule until it fits.

The final schedule is then translated into executable form (by tools), and bandwidth and CPU profile diagrams (also by tools).

5.9.2   Step 2: to determine the priority-based schedule of other tasks in the system.

First, an average bandwidth, and CPU cycle budget, are computed from the profiles in Step 1.

Secondly a complete list of tasks, deadlines, periods is generated and RMA analysis (rate monotonic analysis) done.

If the system can meet all its deadlines and CPU cycles remaining are acceptable to execute remaining non-hard-real-time tasks, we are done. If not, some initial assumptions or parameters must be relaxed and the process reiterated beginning at either Step 1 or Step 2.

6.   CONCLUSION

Platforms are proving to be an effective strategy to cope with ever increasing product complexity, and to facilitate both Hardware and Software reuse. The on chip SoC communication infrastructure decisions are critical to effectiveness on hardware reuse. Proper infrastructure decisions can make the design timing closure friendly and derivative design friendly, as well as decoupling the Device IP library assets from the evolution of the SoC communication infrastructure. Similarly the Software architecture is critical to effectiveness in software reuse. Proper software architecture can provide the flexibility that is key to cope with the diversity of simultaneous sources and sinks of video/audio data with the need to support multiple simultaneous players, recorders, transcoders and rendering options.

The Philips Nexperia Digital Video Platform embodies state of the art approaches to SoC hardware and software infrastructure and is architected for maximum flexibility of creation of cost effective products. The Nexperia-DVP approach recognizes that current and future SoC functionality can only be exercised via software, and therefore Philips Semiconductors takes a system-level approach to SoCs, delivering a software platform and total system solutions. This dramatic shift from silicon to software and systems is well under way, and our Nexperia platform approach is proving to be one of most effective ways to meet today's SoC business challenges.

REFERENCES

1.	http://www.semiconductors.philips.com/platforms/nexperia/solutions/
2.	Donald A. Norman, *The Invisible Computer*, MIT Press, 1998.
3.	Annabelle Gawer and Michael A. Cusumano, *Platform Leadership*, Harvard Business School Publishing, 2002.
4.	http://www.semiconductors.philips.com/platforms/nexperia/
5.	http://www.semiconductors.philips.com/platforms/nexperia/solutions/solutions/viper/
6.	http://www.mips.com/
7.	http://www.trimedia.com/

8.   Rathnam and G. Slavenburg, "An architectural overview of the programmable multimedia processor TM 1", *Proceedings of Compco 1996*, pp. 319-326, IEEE CS Press, 1996.

9.   http://www.vsi.org/

10. Bart Vermeulen, Steven Oostdijk, Frank Bouwman, "Test and Debug Strategy of the PNX8525 Nexperia Digital Video Platform System Chip", *Proceedings of the International Test Conference 2001*, IEEE Computer Society Press, pp. 121-130, 2001.

11. Santanu Dutta, Rune Jensen, and Alf Rieckmann, "Viper: A Multiprocessor SOC for Advanced Set-Top Box and Digital TV Systems," *IEEE Design & Test of Computers*, September–October 2001.

12. Lakos, *Large Scale C++ SW Design,* Addison-Wesley, July 1996.

13. M.H. Klein et al, *A Practitioners' Handbook for Real-Time Analysis: Guide to Rate Monotonic Analysis for Real-Time System*s, Kluwer Academic Publishers, 1993.

Chapter 5

THE TI OMAP™ PLATFORM APPROACH TO SOC

Peter Cumming
Texas Instruments

Abstract: Platform-based design of SoC, as practiced by Texas Instruments, has two key characteristics: platforms are defined hierarchically and software plays as critical a role as hardware. We illustrate these points using the TI OMAP™ platform as an example. Development of new platform family members requires a number of system-level design processes to be carried out. Multiprocessor platforms need a particular focus on SW architectures. We conclude with a detailed description of the TI Wireless SoC platform.

Key words: OMAP, SOC, Platform

1. INTRODUCTION

In this chapter we will use Texas Instruments' OMAP™ platform to illustrate the hierarchical nature of platforms as well as the critical role played by software as well as hardware in platform based design and system development.

OMAP products are combinations of hardware and software allowing multimedia capabilities to be included in 2.5G and 3G wireless handsets and Personal Digital Assistants (PDAs). These capabilities include video messaging, web browsing, video conferencing, games, mobile commerce and many other computationally intensive tasks. End users will demand these new performance and security intensive services while continuing to insist on lightweight, small form factor terminals with longer battery life. To meet these needs, OMAP applications processors and modem plus application devices integrate general purpose computing engines, hardware accelerators for key applications, memory and a rich set of peripherals and interfaces.

As the performance requirements on our devices increase, we must regularly introduce new features to meet the end-consumer's expectations while maintaining cost sensitivity. These factors oblige us to have a highly optimised design flow from TI's architecture team through to our customers' products appearing on the market. Furthermore, to meet the cost goals, we must always be at the leading edge of process technology for migrations (as products ramp in volume) as well as for new products.

The OMAP platform is thus firmly in what has been described as 'the leading edge wedge' where performance, power, cost and time to market are all critical design parameters. From its inception, the OMAP platform was designed to optimise our customer's time to market. However, the first generation devices were developed with a traditional design flow with what has been referred to as 'opportunistic reuse' [1]: we based the development as much as possible on previous designs and made changes as necessary. This pragmatic approach worked well in the very early days of the application processor market but does not support 'leading edge wedge' designs. We have therefore migrated to a more structured approach, known as 'systematic reuse' or an SoC Platform. The remainder of this chapter discusses these issues in more detail.

2.   HIERARCHY OF PLATFORMS

We define a platform as a packaged capability used in subsequent stages of development to reduce development costs. Hence platforms have several characteristics:
- Between silicon and systems many platforms may be developed and used in subsequent stages of a development
- Platforms are valuable due to the notion of reuse, bringing economies in development effort and confidence through wide deployment
- They include hardware, software, assemblies and tools, as appropriate.

By this definition there are some key low level platforms:
- Transistor and ASIC libraries are the lowest level hardware platforms
- Instruction set architecture and associated assembly language tools are the lowest level in software followed by high level languages (and tools plus basic runtime libraries).

In this chapter we consider only platforms above these well understood building blocks.

Figure 5-1 illustrates the levels we will discuss in relation to the OMAP platform.

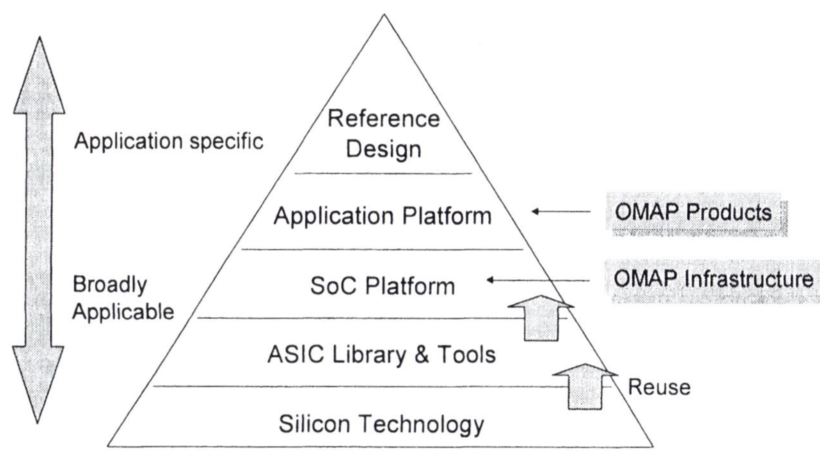

Figure 5-1. Hierarchy of platforms in OMAP processors

The uppermost level, the reference design, is a system platform. Users of this platform can rapidly apply it with minimal engineering effort to build a product. This level of platform brings together multiple heterogeneous integrated circuits; examples in our domain are the baseband processor, application processor, power management and RF. From a software perspective the platform includes code that controls all aspects of the system from device drivers up to the user interface. At TI, this system platform, developed by our reference design team, is a key customer of the OMAP product development team:

- It is a key capability for TI's customers who need to rapid develop products
- It provides us with an internal source of system level expertise that complements the expertise of our customers and the OMAP team
- It is basis of a complete offering of devices and software (power management, baseband modem, RF)

The next level of the hierarchy is the OMAP product – a *full application platform* [2]. The vast majority of embedded systems are based on one or more full application platforms: a standard piece of hardware (one or more chips), typically including a processor and peripherals along with associated low level software and a development environment. Such platforms amortise the spiralling development costs of deep sub-micron ASICs and are therefore the best choice for the many applications that need near-optimal power, performance or area.

The OMAP hardware and software are themselves built using an *SoC platform* infrastructure. An SoC platform consists of, at least, a library of reusable hardware modules (components) and an architecture for their interconnection (rules determining legal collections of components in a product and their interconnection).

3.   THE OMAP HARDWARE-SOFTWARE PLATFORM

The OMAP product range consists of several families of devices targeting different markets. At the time of writing, the range includes: application processors for rich multimedia 3G terminals such as the OMAP1510 and OMAP1610 devices [3], shown in Figures 5-2 and 5-3; to a family integrating a 2.5G modem with low cost application processing OMAP710 and OMAP730 devices.

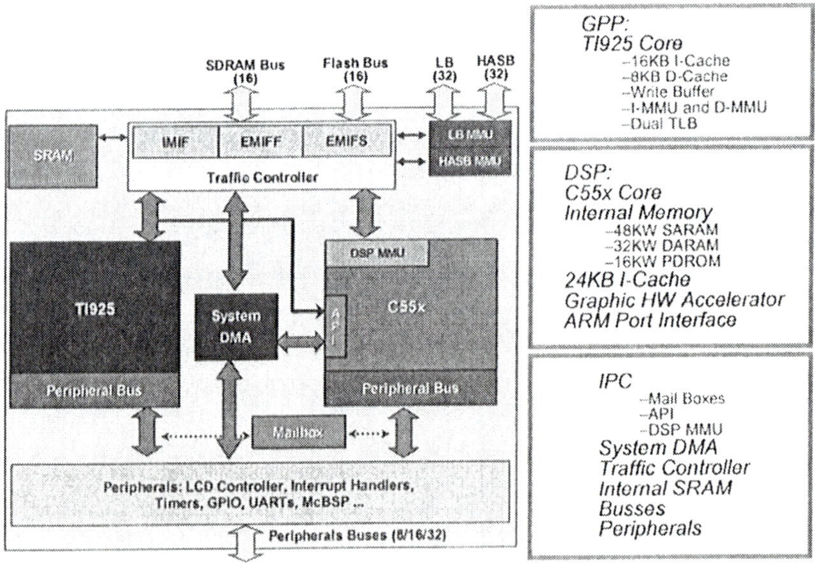

Figure 5-2. OMAP1510 internal architecture

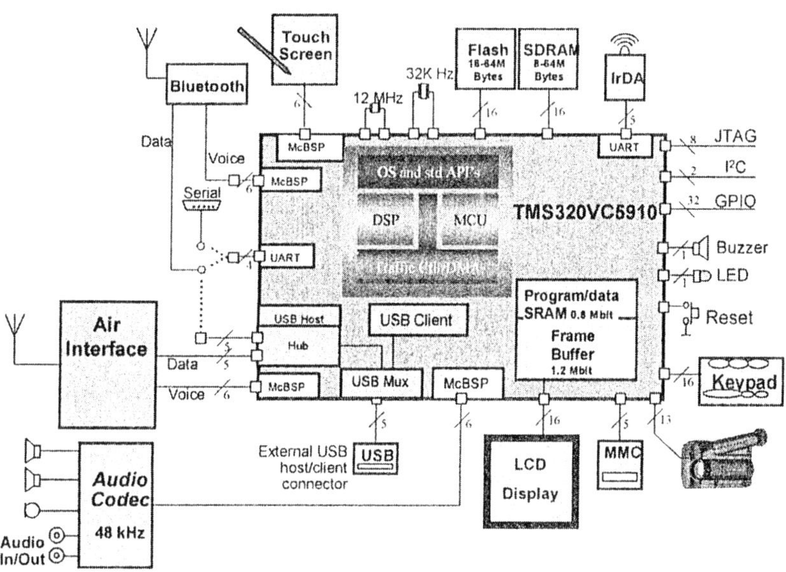

Figure 5-3. A smartphone application of the OMAP1510/5910 device

The strength of such a 'family of processors' is that mobile device manufacturers, software developers and TI's internal development teams can make extensive reuse of development effort across the different products. This applies at many levels such as:
- the devices use common peripherals and memory interfaces and have similar support for interfacing to other components of a mobile phone or PDA, benefiting customer hardware and software development teams
- a common development environment ensures that applications can be shared between devices
- a single low level software framework ensures that component developers can write one set of code and use it on multiple products
- a single SoC Platform can be built and rapidly leveraged to produce products for a specific need

Some of these benefits become stronger as other application chip suppliers adopt the same interfaces to external HW and SW and the Open Mobile Application Processor Interface standard (OMAPI[SM]) [4], founded by TI and ST Microelectronics, is bringing these advantages. We see such reuse and standardization between different platform devices and vendors as an important part of the evolution of this approach.

Once a robust SoC Platform has been established, development of a new family member can be derived primarily from system engineering and focus in this area is critical to the success in the wireless market. This is an area that requires significant investment since it is by definition multi-disciplinary. Some of the key components of this work are listed below.

Application engineering: first and foremost, we aim to build the products that customers ask for. As a leading supplier to the wireless handset market, we have excellent contacts with the development teams at key customers. Understanding and meeting their needs, from the highest to lowest levels of the product is the keystone of OMAP product definition.

Reference designs: teams using the products and thinking about future products give our chip definition teams essential insight into the issues our customers face. While inputs from the industry leaders give the best lead on future requirements, an in-house source of information can provide data or explanations that would otherwise be missed because of confidentiality or other priorities.

Software architecture and development: OMAP processors arrived early in the application processor marketplace and were conceived from day one as a mix of hardware plus in house and third party software. The knowledge gained through the in-house software development and close relationships with major OS vendors (such as Symbian, Microsoft and PalmSource) and 3rd party OMAP developers (ActionEngine, Bitflash, Certicom, Comverse, Hi Corporation, Ideaworks3D, Microsoft Windows Media, PacketVideo, RealNetworks, SafeNet, SpeechWorks) fed into definition of subsequent products both through the TI software teams and third party collaboration.

Performance evaluation: large SoCs need increasingly sophisticated performance analysis techniques. In the case of an open yet targeted platform such as the OMAP platform this consists of several phases:
- Estimation of the workload. The total delay through chip and telephone development, approval and roll-out can lead to a 3 year gap between the start of a development and real deployment in the field. The dynamic nature of the 3G market makes workload prediction a critical first step. This is exacerbated by the fact that we develop products for consumer, battery powered applications with stringent cost and energy budgets, preventing the blind application of high performance processor techniques that typically come with a cost in both these areas. Instead, we predict the workload and build a judicious mix of general purpose and application-tuned hardware.

- Architecture exploration. For high level tradeoffs, it is important to have a model that runs fast and is quickly changed. One example of such tradeoffs is to understand tradeoffs of cache size (miss rate), bus architecture / latency and main memory bandwidth. We use a PetriNet model for these evaluations, driving the processor models with statistics gathered from applications running on previous generations of our products.
- Architecture tuning. Once the basic architecture is defined, there are typically many parameters to be tuned – examples include details of arbitration policies and depth of buffering. For this type of evaluation, a cycle by cycle simulation model is indispensable and we develop initial 'performance only' transaction level models. These simple (so quick to change) models are driven from traces or statistics and hence need only model the level of functionality needed to estimate the performance. An SDRAM controller, for example, need only model the bank tracking logic (to return the correct latency based on whether or not a new row needs to be opened) but need not include a model of the memory itself. As the specification firms up, we can use these performance models as a starting point for fully functional, cycle count accurate models (see below).
- Performance verification. During the definition of a product we specify key parameters, then continuously track these metrics during development. Chip area is one obvious parameter but energy and time consumed for various low and high level operations (cache refill, video decode) are also included.
- Silicon evaluation. When the chip and its associated software are available we can verify the workload assumptions and the correctness of the various models and approximations we made to make the design task possible (for example, we cannot possibly exercise all the workload scenarios on the RTL simulator so we specify parameters that we believe capture the requirements of the scenario). We can also extract extensive trace and statistical (profile) data to support future specification work.

4.   MULTI-PROCESSOR SOFTWARE ARCHITECTURE

One aspect of the OMAP platform that deserves particular discussion is the application and implication of the heterogeneous multi-processor architecture used in the OMAP1510, OMAP1610 and OMAP5910 devices. This architecture is used since, the energy and area constraints of wireless terminals makes a powerful but expensive (in energy and die area) uni-processor unattractive. In addition, partitioning the different streams such as the user interface and the (soft) real time video processing can ease software development and integration – provided the interface mechanisms are well defined and efficient.

In many application areas, it is common to add hardwired logic to address the above concerns. However, this approach is not appropriate when target workloads are unclear and fast time to market, for both silicon and differentiated mobile products, is a critical issue. For these reasons, OMAP1510 and OMAP5910 devices, which appeared in products in 2002, were based on a heterogeneous multi-processing core – a TI-enhanced ARM9 well suited to general purpose tasks and the OS / user interface processing and a TI c55x DSP offering high efficiency for the real time signal processing tasks. As an example, Figure 5-4 shows the division of work between different processors for a video application.

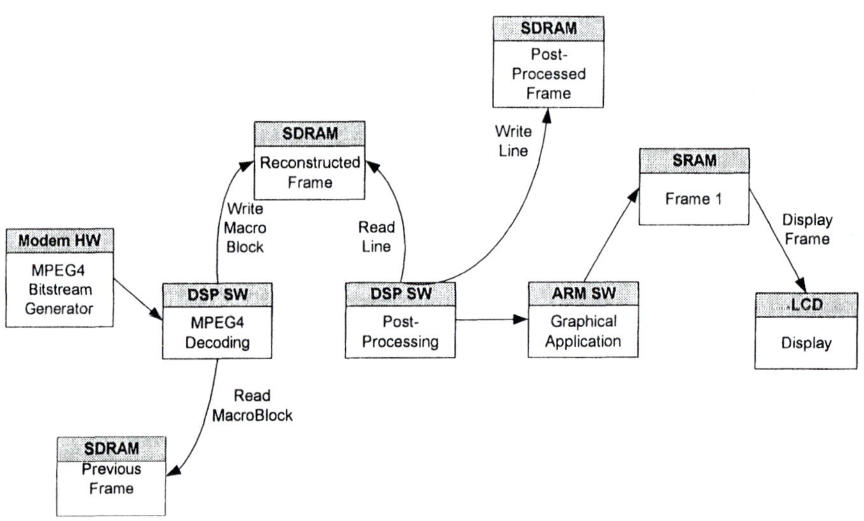

Figure 5-4. Basic data flow for video decode

This approach has proved very valid, with several new applications emerging that can be efficiently performed with the aid of the DSP but which would have been impractical because of energy or available performance (mips) concerns on a single or dual microcontroller (MCU) based device. However, this hardware architecture requires a matching software approach and this is a major focus of the OMAP family.

The approach we took was based on the work done by Spectron (now TI Santa Barbara). It consists of 3 basic elements:

- A well defined set of application programming interfaces (APIs) in the 'high level' OS running on the MCU ('high level' here refers to the ability to manage a virtualized memory space and provide a sophisticated user interface)
- System software linking MCU applications to DSP components such as video and audio codecs
- A well defined standard for DSP components, allowing them to be easily encapsulated in the OMAP framework

We worked with OS vendors to ensure that the necessary APIs were in place, initially focusing on multimedia. This enabled TI and its 3rd party OMAP Developer Network to deliver the power efficiency and performance of a codec running on an optimized processor to the large community of application developers for each of the high level operating systems.

The DSP component standard already existed and was in use by many of TI's extensive network of DSP 3rd parties. This standard (known as TMS320™ Algorithm Standard, or eXpressDSP), requires that components are not only callable from an arbitrary C program but are also well behaved in terms of memory use and access to hardware resources. This allows the components to be linked through *socket nodes* to the multiprocessor communication engine known as *DSP/BIOS Bridge*. DSP/BIOS Bridge can be decomposed into a basic driver layered with code that manages the DSP – the basic architecture is shown in Figure 5-5.

DSP/BIOS Bridge, the socket nodes and the implementation of the MCU OS APIs (known as *Gateway Components*) make up the OMAP multi-processor system software. This architecture supports:

- Dynamic task creation and destruction on the DSP, managed from the MCU, as shown in Figure 5-6
- MCU interrogation of resource (memory, processing power) availability on the DSP
- MCU and DSP cooperation for memory allocation
- Data streaming between MCU and DSP tasks
- Basic IPC constructs between MCU and DSP tasks

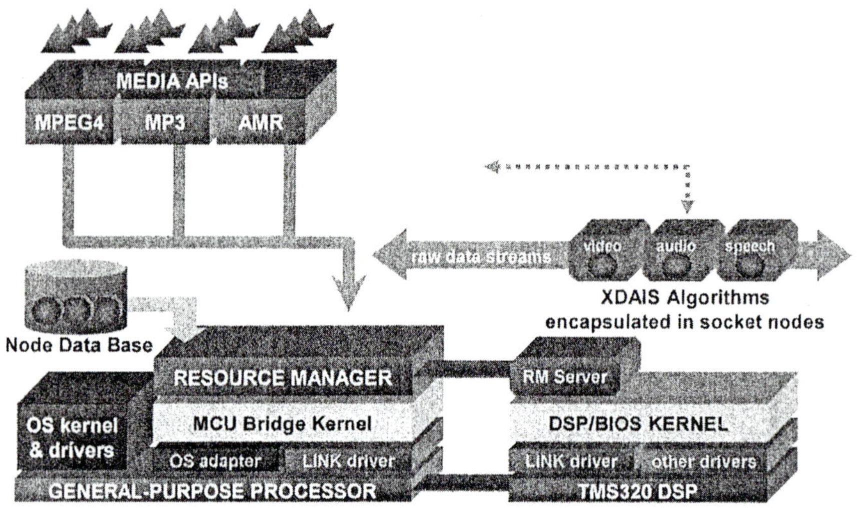

Figure 5-5. OMAP multi-processor software architecture

These features combine to allow DSP components to be developed using classic DSP tools in an environment familiar to DSP programmers. These can then be called transparently from the abstract frameworks provided by high level operating systems. Critically, this allows the dedicated hardware co-processors (ISA extensions) tightly coupled to the DSP core to be fully exploited. This is achieved without limiting the flexibility of application mix on the MCU side and retaining the benefits of a programmable but optimized processor on the DSP.

The typical flow for a DSP enabled MCU application is:

1. Select and attach to a DSP
2. Allocate and connect DSP nodes
3. Create the nodes on the DSP
4. Start the DSP nodes running
5. Stream data to/from DSP nodes
6. Exchange messages with DSP nodes
7. Terminate DSP node execution
8. Delete the DSP nodes
9. Detach from the DSP

The results of this approach can be seen from application level benchmarking that shows DSP enabled OMAP processors (currently OMAP1510, OMAP1610 and OMAP5910 devices) to reduce energy by up to 75% compared with uni-processor approaches while allowing additional optimized functions to be added to a product after the silicon has been produced.

•Application calls 'Bridge API to allocate a DSP node from its UUID
• Application calls 'Bridge APIs to connect, create, run, pause, and delete node on the DSP

• Nodes are partitioned into three phases:
    Create - allocates all resources necessary for the node (e.g., data streams, buffers)
    Execute - performs real-time processing
    Delete - closes streams, releases all resources allocated in the create phase
• Node context passed between phases
• *Execute* phase runs as a real-time task
• Each phase may be dynamically loaded

Node Data Base

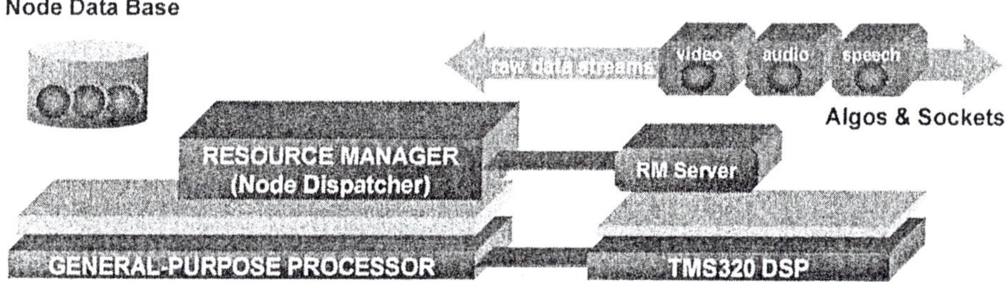

Figure 5-6. Inter-processor communication in the OMAP platform

5. THE TI WIRELESS SOC PLATFORM

OMAP products are built from the TI Wireless SoC platform. In this section we present some of the conclusions we have drawn from our experience prior to and during OMAP processor development and the resulting platform.

The principal driver for this approach to chip design is the productivity of the application platform (chip) design team and the time to market they can achieve. This drives the systematic reuse of IP modules as SoC platform components, but also mandates that these components be simple to connect together – in practice that they all share a common look and feel – and that the resulting assembly is simple to verify. The verification problem illustrates the key aspect of such an approach: it is essential that each module is fully verified before it is instantiated in the assembly or chip and that only the correctness of its instantiation needs to be verified at the assembly or chip level. This principle, known as decoupling, is also critical in the logical and timing design domain as well as in verification.

Decoupling requires that the interface(s) between each IP block and the remainder of the system are well defined. We call these interface(s) socket(s) since they include many aspects of the behaviour of the module. This concept of a socket is related to that proposed by the Virtual Socket Interface Alliance (VSIA) [5] although our current definition is limited to the 'inward' (processor or bus) interface, similar to VSIA's VCI (Virtual Component Interface). Our chosen socket is based on the Open Core Protocol (OCP) [6], primarily for the dataflow interface (or interface family). This aspect of the socket greatly impacts the time it takes to complete and verify the initial design as well as the evolutions that follow (is the interface well specified - including timing? Does it match the IP block requirements? Does it reduce the chances of finding bugs at chip create? Is it robust against changes in interconnect and other architectural parameters? Can the component be easily migrated into other platforms with alternative interfaces? Is it widely supported in the IP and EDA industries?). We chose OCP over VCI since we felt that OCP was the natural evolution of the VCI work and, with the formation of OCP-IP, could form an open, community owned, industry standard for TI, our customers and our IP providers.

For the designer of a component and for the engineer(s) assembling these components into application platforms or custom chips, it is essential that the SoC platform standardises on many parameters beyond the dataflow interface. The following sections discuss aspects of the socket: hardware, software and integration into a platform.

5.1   The ideal socket

Sockets are differentiated from interfaces by their goal of completeness so it is natural to start a description of an ideal socket by enumerating some critical parts of its scope.

5.1.1 Dataflow

All aspects of the dataflow of a module are the basis of a socket interface, but are also covered by more conventional interfaces such as AHB-lite [7] and hence, will not be discussed here.

Several important aspects of dataflow are endianism, address granularity and device width. These areas often form a particular challenge for integrators and are a very common source of errors when less experienced engineers are responsible for related aspects of the design. We have found this to be an important area where very clear guidelines and standards are required.

5.1.2 Clock cycle

The goal of a socket is to decouple module design from integration and allow module designers to focus on their task without adding the concerns of the chip integrator. The chip integrator is, however, critically concerned by the clocking of the devices he or she uses:
- The hardware level interfaces must connect with clean clock domain crossing to avoid risky and latency-inducing synchronization logic.
- The global clocking scheme providing clocks to both interconnect and the various modules must be as simple as possible to minimize power.
- The system level performance, which depends critically on component clock speeds and interface latencies, must be adequate.
- The final chip level timing closure must be achievable and ideally straightforward.

These considerations mean that a true socket, just like its board-level equivalent, must include budgets for both clock cycle and interface signal timings.

Of course, in the SoC arena, there are often hierarchies of busses as well as multiple target technologies (from different silicon vendors, but also through technology migrations). The ideal socket therefore, includes requirements on timings for a reference technology, but also includes 'documentation' that gives details of actual timings which may be better than those required by the socket definition.

5.1.3 Clocks

It is clear that large SoCs must be based around a simple and robust clocking methodology and that the socket definition must support this. While SoCs will likely migrate to more complex clocking environments such as locally synchronous, globally asynchronous, the socket should remain simple and synchronous.

This goal is not completely straightforward since many modules in a SoC are required to interface to the external world with specific clock rates, often derived from the interface clock (USB, UART, PCI are a few of the many examples).

Hence, synchronization logic is, in general, required between the interconnect and the backend of the module. Two approaches are possible, either the module includes synchronization logic or an additional level of interface abstraction is added as shown in Figure 5-7.

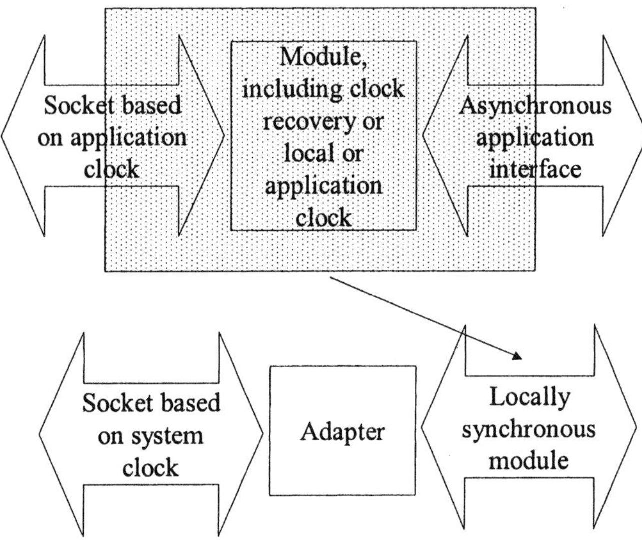

Figure 5-7. Asynchronous peripheral with clock adapter.

The top part of Figure 5-7 shows a module with an application driven clocking regime on the right hand interface and no synchronization logic. Hence, the socket interface timings – whether the module is an initiator or target – are relative to the application clock. The lower part of the figure shows an adapter which takes both application and system clocks as inputs and provides a socket synchronous to the system clock.

In this latter approach, the adapter may be subsumed into the interconnect in the form of an *agent*. This approach is often convenient for the module designer. It is also well suited for integration since in some cases the system clock will be appropriate for the module (for example, a system may be built around a multiple of the PCI clock rate) and the overhead of synchronization logic is not needed.

In other cases, such as links where the clock is derived from the application interface itself, synchronization may be a natural part of the module design. Hence, we allow both strategies – synchronization in the module and synchronization in the interconnect (or a discrete adapter).

5.1.4 Reset

Since the socket concept is intended to ease SoC integration and module verification, it is critical to include the reset protocol in the definition. Hence, our ideal socket defines the signals used to reset the module and the transitions that occur on them (polarity, duration) as well as any restrictions on signals driven by the module during and immediately after reset.

5.1.5 Interrupt and DMA Requests

Signals, protocols (edge or level sensitive) and synchronization (or absence thereof) in interrupt and DMA signalling must be defined. Where possible, we prefer approaches that support an arbitrary number of these event signals and allow the system integrator to determine whether they are to trigger software (interrupt) or hardware (DMA) responses.

This goal of harmonized event signalling suggests the use of an edge, rather than level, based protocol.

5.1.6 Semantics

Any CPU based system must deal with the requirements of different types of access:
- Weak order (or memory semantics) allows reordering and merging of accesses provided *hazards* such as RAW (read after write) and WAW (write after write) are avoided. Data can also be forwarded from a write to a following read.
- Strong order (or IO semantics) enforces that each access completes to its target before following instructions are executed by the processor. This prevents, for example, writes being posted and the CPU released prior to completion of the write at the target. This type of access is used for IO devices where, for example, reads may change the state of the target (such as reading from a FIFO). A second example is the write that clears an interrupt and is closely followed by interrupts being (re)enabled in the processor: the write must complete to the interrupt controller before the interrupt enable instruction is executed.
- Cache coherence and memory barriers are supported by high performance processors. These advanced features are typically absent from current SoCs, but are likely to become more important in the future. Fortunately, they concern only a few types of module and can, hence, be added as needed without requiring a change to the majority of legacy modules. It is sufficient that today's sockets are able to be extended to include these features in the future and OCP meets this requirement.

5.1.7 Scalable Performance

It is natural that we want to use the same socket definition for the full range of modules that we will design or integrate. As mentioned earlier, this implies different clock speeds adapted to different levels of bus hierarchy. More significantly, it may also drive the use of several data bus widths and different levels of sophistication in the interface's dataflow protocol. In particular, a high performance module may support:
–   Split transactions allowing the interconnect (and potentially the module) to efficiently process multiple requests with long or varying latencies.
–   Threading (see OCP specification at [6]), allowing a module to generate or respond to multiple streams (threads) of transactions concurrently, reordering transactions within the threads, but retaining order within each individual thread. This capability is well matched to initiators such as multi-channel DMA controllers and processors and can be exploited by targets such as SDRAM controllers.

The complexity of these features is justified for high bandwidth modules that can benefit from them, but is unacceptable overhead for designers of other modules. Hence, we make such features optional and allow multiple bus widths and clock speeds and the socket definition becomes scalable.

5.1.8 Higher Level Functions

In addition to the signalling level characteristics of the socket described in the previous subsections, it is desirable to include higher level standardization. This may be provided to aid SoC integration or to support driver software's exploration of the hardware on which it is running. The most basic example of such standardization is the allocation of a register – at a fixed offset in a module's address space – that gives basic information about the module. This information may include the vendor, the module identifier and version number.

5.1.9 Extensibility and Flexibility

The above considerations apply to any socket. If, however, we consider a definition that is intended to be used by multiple companies, considerations of extensibility become particularly critical.

Firstly, many companies create IP for integration into their own SoC designs. In doing this, they may wish to benefit from the end-to-end control they have by adding domain specific features to an industry standard socket. Texas Instruments' wireless business unit is an example of such an organization. We design and integrate modules for wireless handsets and hence, one of the areas we naturally focus on is power consumption. We have therefore, defined innovative proprietary extensions to standard interfaces in several areas, including energy management.

For organizations such as ours, the industry standard must include capabilities for user-defined extensions. The ability to define these extensions within the socket is important in areas such as interconnect generation and verification tools. 

Some of our extensions to the interface will be offered to the standards process in the future and we expect to benefit from the availability of additional IP supporting these features as well as from complementary extensions provided by others. The user-defined extensions can hence, be viewed in part as a test bed for additions to the standard.

A second, perhaps more important, area is that of future evolution. The natural goal of standardisation is to enable multiple vendors to design and maintain extensive IP libraries and EDA tool support. Without a socket that ensures continuity, these IP and tool developers will adopt internal abstractions and bridge to sockets based on customer requests. While this approach is feasible, it is clearly not optimal from the perspective of either the developer (who has more work to do) or the integrator (who may suffer area, power or performance penalties).

To avoid this defensive scenario and enable native IP and tool creation, a socket must be:

- Sufficiently flexible that it is a natural fit for the task at hand. Use of the industry standard socket must be at least as easy as the current in-house abstraction.

- Simple to *gasket* (bridge) to current standard and proprietary interfaces (to support interworking of modules and interconnects including legacy elements). Again, bridging must be no more difficult or expensive than with in-house abstractions.

- Based on an extensible foundation – both technically and organizationally – that will support the evolution of the socket and ensure that tools, modules and experience can be transparently reused when the standard evolves. Board level standards such as PCI are good examples of how such an evolution can be orchestrated in a community owned standard. We believe that community ownership is a critical part of socket standardization and are working in the organisation that governs OCP to ensure that the standard meets these goals.

5.1.10 Compliance

One critical aspect of socket use is the ability of an IP provider to verify his implementation and to demonstrate the compliance of the module to the standard. Today, there are few mechanisms for providing independent *certification* of a module and while much simpler, it shares some of the problems of software certification.

Golden vectors for self-certification, coupled with embedded checkers for SoC verification, are the current state of the art. These can be complemented, as at the board level, by 'plug fests': [8] is an indication of how this concept could be applied to the SoC world. The IP and SoC industries have not, and may never, reach the critical mass and level of interoperability that warrant independent compliance checking.

5.2 Platforms

An SoC platform is more than a socket. It includes basic rules for building compatible devices such as:
– CPU instruction set
– Interrupt architecture (single level, multi-level...)
– DMA strategy (centralized vs distributed, programming model)
– Performance parameters (supported socket options)
– Power management capabilities
– Exception handling (error reporting)
– Maximum socket interface speeds

It can be seen from this incomplete list of platform characteristics that the goal is that key pieces of IP can be developed for the platform and seamlessly reused across instances. Hardware blocks such as interconnects and DMA engines and software modules such as interrupt handlers and DMA managers are examples of developments that are not fully specified by the socket – the refinement of the platform definition (Figure 5-8) is needed to ensure interworking.

In TI we have one team producing peripheral and other application specific IP, another that is responsible for the platform development including its generic components such as memory interfaces, DMA and interrupt controllers. These teams supply multiple chip development teams. When we revise a platform or a component it is provided by the platform team to the lead chip development team but is also integrated into the platform team's development environment, known as the *Reference Assembly*. This allows the platform team to sanity check their design independently of multiple and changing chip level requirements.

Figure 5-9 shows how decoupling through sockets and the platform definition combine in design and verification.

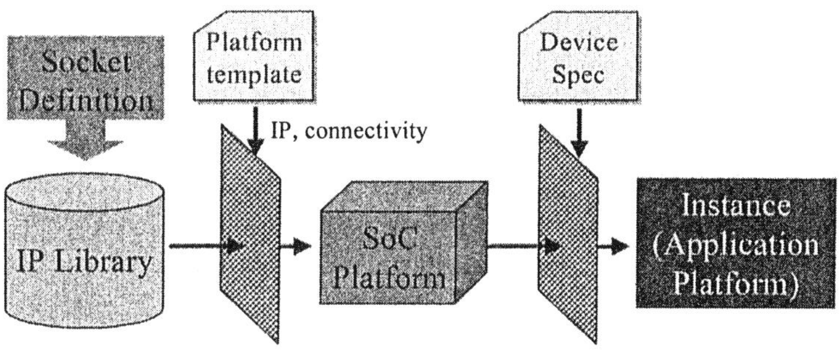

Figure 5-8. Sockets, Platforms and Instances

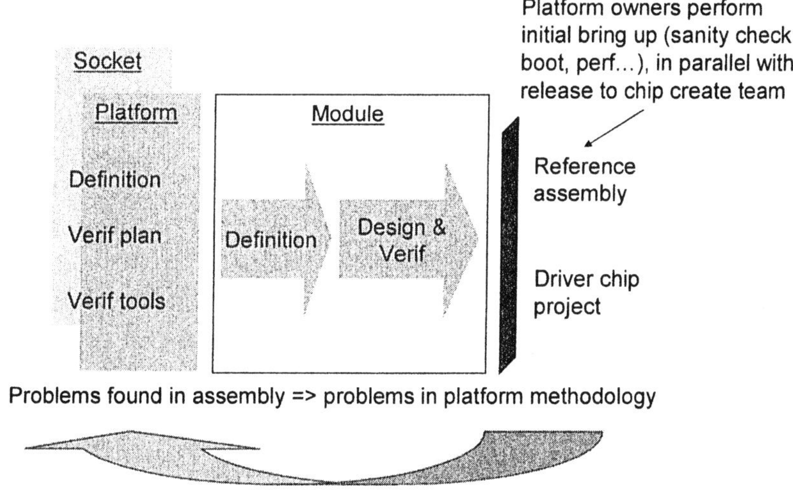

5.3 Software

To support rapid generation of derivatives from a platform it is desirable
that the low level software infrastructure mimics the modularity of the

hardware. Unfortunately, low level software such as drivers is inevitably OS specific since each OS has its own model for the APIs and functionality such software should offer. We are therefore obliged to make some compromises when building a software infrastructure for an SoC platform. The issues to be addressed include:

1. The port of the OS to the basic peripherals
2. The integration of additional (optional) devices
3. Commonality between different devices based on the same SoC platform
4. Reuse between different OSes

TI is typically not responsible for the 'base port' of an OS to a given CPU ISA (in our case, ARM) and hence the first of these items is not a major concern: the driver set for the basic peripherals (timer, watchdog, interrupts etc.) is small compared with the driver set for the rich peripheral mix we include on our devices (point 2 above). Hence we focus on general driver reuse without undue emphasis on the basic OS set.

The previous section touches on some of the issues needed to support the 3rd item – this is a critical enabler for reuse of low level software developed by TI and our customers.

The combination of good coding practices (such as maximum use of constants for addresses and masks) and 'discovery' features in the hardware is also important to enable driver reuse. For critical hardware modules, such as DMA, we provide the ability for the software to query the specific hardware and to alter its behaviour accordingly. This allows a single 'fat' driver to be written for a range of hardware implementations – either in multiple chips or multiple instances of the same module in a single chip, where the parameters of the module may vary from instance to instance. 'Fat' drivers may have been considered overly expensive in early wireless devices but the growth of application and OS code coupled with the availability of bulk storage (such as NAND flash devices holding compressed images of the code) in coarse granularities means that driver code size is no longer a major concern.

The final item – commonality between OSes – combines with a desire to reuse OS specific code across different hardware implementations to lead us to the concept of a *CSL*. The CSL, or Chip Support Library, is the lowest level of software that communicates with the hardware. It provides a minimum abstraction of the full set of services offered by the device in an OS independent way. The driver writers for various operating systems can use this CSL to abstract hardware implementation details.

These techniques all contribute to delivering a complete product: not just silicon but a complete solution including system software linking to optimised multimedia components, not to mention the essential collateral material such as tools, development boards, documentation and support.

5.4 Future Proofing

As mentioned at the beginning of the chapter, our products move rapidly through process technologies and evolve quickly in terms of performance and features, yet we cannot re-invent our SoC Platform for every process node, nor for added features. Hence we will consider the robustness of our approach against some of the likely evolutions in the technology:
- Process scaling (wire, leakage)
- High performance (processor vs. memory speed)
- Additional masters (for example, camera or modem)
- Increasing area, decreasing size (high end, low end, interconnect span)
- New high bandwidth modules and workloads (bigger screens, graphics, importance of heterogeneous, wide interconnect, floorplan aware...)
- Additional, heterogeneous smart accelerators (bridge)

6. CONCLUSION

We have discussed some of the critical issues in enabling platform based design in the leading edge wedge of battery powered consumer products. Delivering the promise of the platform technique to end equipment manufacturers obliges platform providers to address energy, performance and time to market at all levels including:
- A wide product range allowing reuse of hardware and software development and sourcing across end equipment product ranges
- A hardware architecture adapted to problem: matching performance, flexibility and energy requirements
- A software architecture delivering all the benefits of the hardware to the application developer
- An efficient SoC platform comprising hardware and low level software
- A complete and flexible socket allowing hardware to be easily developed, verified and integrated
- SoC platform definition for hardware and software reuse

TI is the market leader in wireless SoCs and with the OMAP product family. Based on our extensive experience we have invested in our OCP based SoC platform to build an industry leading capability that provides a solid basis for evolutions through at least 90nm and 65nm technologies.

REFERENCES

1. Henry Chang, Larry Cooke, Merrill Hunt, Grant Martin, Andy McNelly and Lee Todd, *Surviving the SOC Revolution: A Guide to Platform-Based Design*, Kluwer Academic Publishers, 1999.
2. Alberto Ferrari and Alberto Sangiovanni-Vincentelli, "System Design: Traditional Concepts and New Paradigms", *Proceedings of the 1999 International Conference on Computer Design (ICCD)*, Austin, Oct 1999.
3. Texas Instruments product information available on the web at www.ti.com.
4. OMAPI information available on the web at www.omapi.org.
5. Virtual Socket Interface Alliance, on the web at www.vsi.org.
6. OCPIP, on the web at www.ocpip.org.
7. ARM Limited, on the web at www.arm.com.
8. SOCWorks, on the web at www.socworks.com.

Chapter 6

SOC - THE IBM MICROELECTRONICS APPROACH

Kathleen McGroddy-Goetz, Robert Devins, Michael D. Hale, Mark Kautzman, G. David Roberts, Dwight Sullivan
IBM Microelectronics

Abstract: System-on-a-chip (SoC) designs represent a rapidly growing segment of application-specific integrated circuit (ASIC) and application-specific standard products (ASSP) designs. The Worldwide Design Center (WWDC) in IBM Microelectronics' ASIC organization has been designing SoCs for both internal and external customers since the mid-1990s. The experience gained over the past several years has led to methodology refinements that greatly reduced time-to-market for designs of increasing complexity. The key factors to reducing SoC time to market include standard interconnects, verification methods, the development of reusable platforms, and the enabling of concurrent software development. These factors - and the impact of the methodology on development cycles and resource requirements - will be discussed. Remaining challenges are highlighted in the conclusion.

Key words: Platform-based design, reusable verification

1.   INTRODUCTION

Since the late 1990s, the ASIC marketplace has been increasingly dominated by SoC designs, defined as hardware containing a compute engine, memory, and logic integrated on a single chip [1]. The number of application-specific standard products (ASSPs) being purchased as an alternative to ASICs is also growing rapidly.

The need to get to market quickly is driving the move from traditional ASICs to SoCs and ASSPs. System developers cannot afford to redesign each new generation from scratch. ASSPs can provide a quick path to market once they are available because customer differentiation is done using software. However, an ASSP can take longer to design than an equivalent ASIC because it must be designed and validated for use in many different systems. A viable alternative to ASSPs can be an SoC designed for a single customer for use in multiple applications such as a "Customer Specific Standard Product" or "CSSP." Customers can incorporate their unique logic into a CSSP allowing additional product differentiation over what can be accomplished by software alone. CSSPs can offer reasonable hardware and software development cycles if they are based on an ASSP architecture or "platform" that can be quickly redefined with new custom logic and brought to market rapidly. This places significant requirements on the intellectual property (IP) blocks in the ASSP as well as the methodology.

The IBM WWDC began developing CSSPs in the mid-1990s with several projects targeted at communications and printer controller applications. These designs were all based on the PowerPC 401™ embedded processor. In 1998, a challenging project was undertaken to develop an embedded PowerPC® 405-based SoC for test and measurement applications with Agilent [2]. This SoC contained 26 digital and analog cores provided by IBM, Agilent and several third parties. The complexity of this project and the desire to leverage the experience led to the development of a robust, reusable SoC methodology based on PowerPC, the CoreConnect™ standard buses, and the IBM Blue Logic® core library.

This chapter is organized as follows: Section 1 provides an introduction; Section 2 discusses the importance of standards; Section 3 describes leveraging reusable verification; Section 4 discusses defining a platform; Section 5 describes pre-silicon software development, and Section 6 summarizes the chapter and describes future directions.

2.   THE IMPORTANCE OF STANDARDS

One key enabler of SoC development is the existence of standards for both development and connection of IP blocks. Standard interfaces can be used to facilitate rapid interconnection of IP blocks as well as provide guidelines to customers for their IP design. Having well-defined interfaces can simplify the integration process and promotes design for reuse. It can simplify the development of complex systems and reduce the need to design glue logic that potentially impairs the performance of the system. Standard development practices are needed to map the IP blocks to a target ASIC technology. Standard architectures for SoC chip registers are needed to provide a familiar interface to software developers. Standard development practices are needed to create a reusable framework of software based verification IP, thus minimizing the re-invention of logic verification environments on successive SoC designs. Finally and most importantly is the need to have peer-to-peer standards compatibility and a mechanism to improve these standards over time.

2.1   The CoreConnect Bus Architecture

The CoreConnect bus architecture was developed to be IBM's standard mechanism for connecting IP blocks in SoC designs. It is comprised of three bus types: the processor local bus (PLB), the on-chip peripheral bus (OPB) and device control register (DCR). The PLB bus is a high-performance low-latency interface with overlapped read and write transfer capability. It is typically used to connect high-bandwidth devices such as processors, DMA controllers, and external memory interfaces. The OPB is a secondary on-chip bus intended for lower speed peripheral devices such as serial ports and parallel ports. The DCR bus is a lower speed core configuration bus driven by special PowerPC instructions. The DCR is used to read and write status and configuration registers.

CoreConnect was initially developed for use with the embedded PowerPC 401 processor in 0.5um technology, with the first instance of the PLB running at a maximum frequency of 66 MHz. Today there are four generations of PLB in multiple technologies, running up to 200 MHz and including a crossbar switch version. More than a dozen IP blocks were designed to these initial interfaces in 0.5um technology and the architecture was made an open standard. Having a standard interface allows timing and performance characteristics to be clearly defined, making a derivative SoC chip architecture possible. Making the interface an open standard allows third party IP providers as well as customers to easily design to the standard. Verification of an IP block can be made simpler in this environment.

2.2   Compliance Verification

To ensure compliance with the CoreConnect bus standard, a unit-level test environment is needed for bus protocol checking. Unit-level verification toolkits have been enhanced and reused for successive generations to validate these bus interfaces. These toolkits implement bus functional architecture models of the various CoreConnect masters and slaves. Protocol monitors for each bus, check the correctness of each transaction, as well as generate a log of transactions used. A bus functional language (BFL) allows designers to control master and slave transaction attributes with bus cycle granularity, both statically and randomly. Simultaneous transactions, initiated from multiple masters, allows for the creation of peak bus utilization collision scenarios in a unit-level test environment.

An automated methodology for bus compliance can help to reduce risk to both rookies and seasoned IP block designers and it also helps to provide a basis for peer-to-peer reviews. An example of this automation is the CoreConnect Test Generation (CTG) software. The CTG software generates a bus compliance suite in BFL for particular IP blocks, based on supported bus transaction types selected by users via a graphical user interface. These transaction types are generated from a so called "super table," which describes available transfer types such as read, write, cache and burst, parameters such as byte enables and delays, statistical distribution controls over these parameters, and filtering of non-applicable transactions to the IP block. After the desired options are set up in the super tables, a suite of test cases is generated by CTG to stimulate the device under test and the CoreConnect master and slave architectural models. For determination of test coverage with the test generator, a correlation tool compares the descriptions in the super table with the simulation protocol monitor transaction logs. Future directions in automation include toolkit enhancements for dynamic on the fly transaction control of CoreConnect models and non-CoreConnect chip interfaces, interfaces to hardware verification languages for control of the bus functional models and coverage, and use of formal validation methods as an alternative to traditional simulation. Once the new IP block is designed and functionally validated at the unit level, system-level verification is focused on correct interconnection of IP blocks.

Formal assertion based verification techniques represent an opportunity to augment traditional simulation based verification at both the IP block unit and chip levels, as well as providing a bridge to formal validation analysis of a design. An example of this is the Sugar language. With assertions coded in Sugar, a block of assertions can be developed for a particular CoreConnect bus to check for protocol correctness. At the unit level, assumptions of IP block designers can be documented in those assertions, which may not be necessarily apparent from the IP block paper specification document. These assumptions at the unit level can be quickly tested at the chip level when neighboring IP blocks may violate these assumptions. Coding standards will need to be put in place to successfully facilitate reuse of assertion-based methods for IP blocks from the unit to the chip level.

2.3   Compatibility and Best Practices

The CoreConnect buses have maintained backward compatibility while increasing performance through successive processor and technology generations. This backward compatibility has maximized reuse of legacy IP and allowed a migration path to new generations with minimal redesign required. Many of the IP blocks that were developed initially for use in the 0.5um generation are still in use today. They have been migrated through successive technologies including 0.25um, 0.18um, 0.13um, and 0.09 um for use in systems powered by newer generations of embedded PowerPC processors. Although some minor modifications are required occasionally and new deliverables added, the initial investment in developing reusable IP blocks has proven extremely worthwhile.

In addition to CoreConnect, standard development practices are needed to map IP blocks to a target ASIC technology. Requirements documents and methodologies define the ASIC development practices and deliverables. Backing examples and templates assist in enforcing the written requirements. Standards can be as simple as filename or module instance names, or as complex as spelling out the format of static timing assertions and a method of scannable latch test insertion and coverage. There are three main types of IP blocks defined by the Virtual Socket Interface Alliance (VSIA) taxonomy: soft, hard, and firm cores. So-called "firm cores" are provided in a technology specific netlist. Firm cores are designed in RTL and then synthesized and timed to restraints defined by the core development requirements. Hard cores have a pre-determined fixed physical layout and are wired and timed. Hard cores, such as the microprocessor, also need an encrypted "black box" RTL or netlist cycle accurate simulation model of the IP block. Soft cores are typically delivered as RTL, enabling better simulation performance and facilitating debug, but the ASIC technology synthesis is left as an exercise to the end user. A mechanism to amend and improve the IP block interfaces to the ASIC technology is needed to allow evolution of the IP block deliverables over time.

A standard architecture is needed for SoC chip control registers. This enables portability of system software, reuse of chip architecture documentation in derivative designs, and can help reduce the learning curve for users familiar with existing SoC designs. The implementation of this interface includes a system address map, a read/write register access mechanism, and a standard register bit definition for each chip control function. Some examples include clock generation control for the microprocessor and peripherals, power management control, and core specific controls such as burst lengths. Derivative designs can omit the control registers from their chip implementation if the IP blocks behind them are not present, such as a PCI or Ethernet subsystem. These registers also allow for control of common clocking circuit architecture for both integer and non-integer CPU: PLB clock ratios for PowerPC microprocessor based SoCs.

2.4   Making Standards Successful

System verification software needs standard development practices and a documented methodology. SoCs today at IBM are simulated and verified at the system level with a microprocessor-based software known as the Test Operating System, or TOS. For verification software to be reusable, coding standards must be put in place as an integral part of the development methodology. Verification software test scenarios need to be partitioned into core or functional sub-system based areas so they can be easily added to or subtracted from a derivative SoC design. Verification software should be easily parameterized for interoperability with different address maps, interrupt configuration, and sharing of system resources such as memory regions or shared I/O pins. Written standards by themselves do not make code reusable. Code generation tools or template generators are one means of creating a consistent framework for software tests. In the end, code and test coverage reviews with peers are often needed to ensure completeness and quality.

One of the most important underlying components in making standards successful in an SoC methodology is the need for peer-to-peer communication. In the end, it is the human factor that closes the gap in meeting standards, developing IP blocks and designing SoCs for fabrication. Sometimes this can be one of the most challenging aspects of development, especially when design teams are in geographically different locations and have different views and interpretations of the standards. To span the geographic gap, it is imperative to have development management and several key engineers embrace and implement the standards in all facets of their development. A compatibility review board or code review board is essential for determining if requirements are met, both early in the design phase and after the final deliverables have been completed. These review boards typically consist of the IP Block developer or code developer and the team of experts familiar with the details of the standard interfaces. Early in an ASIC technology there needs to be a crisp definition of the IP Block requirements. Sometimes, standards need to be amended and improved through learning or changes in the base ASIC technology. To enable change, both a process and periodic human communication are needed.

3.   LEVERAGING REUSABLE VERIFICATION

One of the more significant problems plaguing development of a SoC is verification. The use of pre-developed discrete intellectual property blocks gives design engineers an unfair advantage over their verification counterparts. With a suitable IP collection, designers are able to stitch together multi-million gate designs quickly, which then need to be proven functionally correct. The daunting task of providing this proof then falls into the hands of the verification community.

Not only must the correctness of the design be verified, but the whole design also must be proven to be suitable for the target application. To be able to perform this work in a timely manner, a process that leverages previous verification work must be developed and implemented. The resultant verification effort must be able to perform complex scenario-based transactions on the full chip and be able to control the surrounding simulation environment (for stimulus and response synchronization). Buses and design resources (cores, memories, etc.) must be utilized concurrently just as they will be used when the design is realized.

IBM's WWDC has developed the Test Operating System (TOS) methodology to meet these needs. The TOS methodology consists of a minimal software kernel, a C-language API, standardized development and quality rules, a developer tool suite, and complete documentation [3].

At the center of the TOS methodology is the kernel. This minimal "operating system" was developed in C and provides the complete set of facilities required to boot the design and to run test applications. The kernel handles everything from processor initialization to boot vector setup, interrupt handling, task scheduling, memory management and resource allocation. TOS is designed so the kernel is small enough to be booted during simulation, something not typical of traditional real time operating systems (RTOS). The result is then compiled with associated tests and loaded into a memory image for simulation. It also can be directed to target other environments such as an instruction set simulator (ISS) or even real hardware, allowing reproduction of events across all platforms.

The TOS operates through a messaging mechanism, which means that extremely complex situations are possible. Activities such as multitasking, multiprocessor support, interrupt callbacks, hardware contention, memory allocation, mutual exclusion, error injection, and external stimulus synchronization all occur with complete test case transparency.

Coupled with this kernel is the TOS API. Documentation, on how to develop test cases that operate within the framework, is provided. Test cases are written in standard ANSI C and only are allowed to access library functions provided by the kernel. This lack of dependence on other libraries provides extreme portability across a huge variety of verification targets. The processor in the design is often replaced with a bus functional model (BFM) which can yield a 20X improvement in event simulation. In this situation, the kernel is compiled to be a process, which runs on the simulating workstation in lock-step to the simulation. When simulation input or output is required, an access is executed via Verilog PLI to the BFM. All of this is performed transparently to the test case simply by recompiling and redefining the operation of the underlying TOS API. Test cases work across all targets.

To further modularize verification, each core not only has TOS test cases, but also has a device driver. Device drivers provide a hardware abstraction layer (HAL) for TOS library functions, and provide both specialty operations and the register programming interface for any given core.

Device driver requirements and formats are specified in the TOS documentation. When the TOS needs to exercise the hardware, it will only use these device drivers to do so.

All TOS test cases and device drivers are subject to a series of quality metrics and independent reviews. The purpose of these quality checks is to ensure that all core connectivity is exercised and that typical true operation of the core is also seen during simulation. Once these checks are complete, the code is checked into the ASIC library along with the core itself and made available to customers both inside and outside of IBM. When customers obtain a core, they automatically obtain the system-level verification module, which goes along with that core.

Typically when developing a new design, the cores and TOS code are taken out of the library. While designers are busy putting the chip together, the verification team quickly creates the chip-level exerciser. This chip-level exerciser is the main C application, which runs under the TOS kernel. It contains all of the chip specific information, such as core offsets, bus locations, interrupt numbers, and reset senses. The exerciser is also where verification engineers turn tests on or off and indicate whether the tests need to be run alone or concurrently with other tests in the suite. TOS integration occurs in a very short period of time and often predates hardware integration.

The following diagram illustrates the major block diagram of TOS operations:

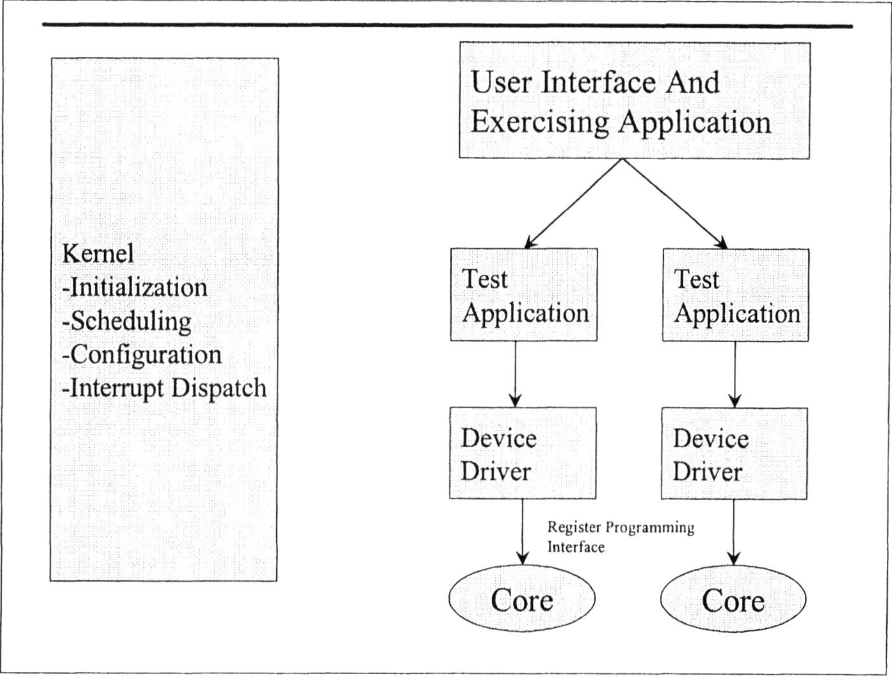

Figure 6-1. TOS Software Structure

Since so much of TOS is based upon standards and documented coding guidelines, several tools have been developed to ease the creation of TOS components. The developer's tool suite consists of device driver, application (test case), and system description (chip-level exerciser) generators. These generators can take condensed descriptions and create hundreds of lines of TOS code. This code may either be complete (as in the case of the exerciser generators) or be reconfigurable frameworks waiting only for the code body (as in the case of the test case generator).

The use of TOS as the primary system-level verification methodology has led to huge reductions in both the number of engineers needed and the effort required verifying a complex SoC design. The SoC verification methodology now consists of developing a verification plan based on the existing reusable test cases, determining whether any chip-specific tests must be written, regression maintenance, and debugging any observed failures.

4.   DEFINING A PLATFORM

SoC integration was greatly simplified as IP standard interfaces came to fruition. However, the task was still time consuming. SoC verification assumed the IP cores were functionally correct, so the interconnectivity verification was left to the SoC design team. SoC design teams had to understand a great deal of IP functionality to be able to verify its connectivity. Multiple design teams were implementing solutions to the same problem, leaving less time available to concentrate on the SoC's design and verification. A SoC reuse methodology had to be defined to improve time-to-market and improve quality.

Initial efforts in developing a reusable SoC methodology were focused on the idea of starting with individual pieces of IP as the basic building blocks. Proceeding this way drove significant new requirements onto the IP developers and was inefficient from a chip integration perspective. The major concern with a building block approach is that the same cores usually get stitched together over and over in different designs, thus increasing the chance of simple user errors. Lessons learned from the initial development efforts indicated two possible approaches:

– Bus centric reference platform
– Standard product platform

The standard product platform was considered too rigid as a starting point for derivative designs, so a bus centric platform was chosen. Information was gathered from several previous designs to develop a bus centric approach. Common IP and architecture from previous designs were selected for the base SoC.

The first platform, shown in Figure 6-2, consisted of a PowerPC 405 processor, standard PLB and OPB CoreConnect buses and the IP common to previous design efforts. These included memory controllers and an external bus controller core on the processor local bus (PLB) as well as the memory access layer core (MAL) that interfaced to a 10/100 Ethernet macro. A bridge to the on-chip peripheral bus (OPB) was included and enabled inclusion of lower speed peripherals including two serial interfaces, an IIC, a general-purpose timer core and general purpose I/O. The end result was then verified to be correct.

Having a common, preverified SoC set was not enough; thus, further efforts were taken to:

– Ease integrating application specific IP into the reference design
– Ensure maximal configuration capability
– Create an environment for simulation
– Create an environment for verification
– Offer a sample flow for realization (synthesis, timing, floorplanning, etc.)

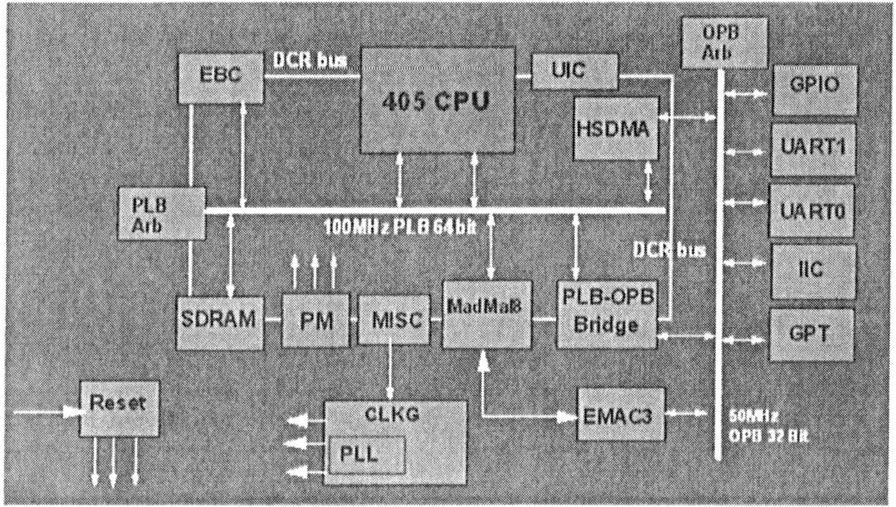

Figure 6-2. PowerPC® 405 Platform Based Design

Integration of application specific IP was the differentiating feature per SoC in a bus centric platform methodology. Defining a methodology to add or remove IP from standard interfaces seemed trivial but the implementation was usually very awkward and manually intensive. The current methodology required the SoC designer to edit the top-level interconnect files. This method assumed more risk in having the preverified platform inadvertently modified to an unusable state. To minimize the risk, the platform had specific address slots or ports available on its respective buses. The platform also used Verilog's "`define" to be able to appropriately remove unneeded cores. Still, the challenge of correctly instantiating, connecting and verifying new IP was the sole responsibility of the SoC designer.

With a preverified platform, the SoC design team was allowed to concentrate on the details of the changes made to a known "golden" model, which essentially reduced the number of variables changing at once. The verification environment was flexible enough to allow address changes of individual IP without affecting the individual tests. Over time, a repository of IP verification code was developed; even cores that were not part of the platform offering could easily be integrated and verified.

4.1 Platform versus Platform Based Design

Up to now, only the platform proper has been discussed. Platform based design must go a step further and enclose the platform itself into a product

like offering. Platform based design can be summarized as a predefined well documented environment for the design flow. Having an environment defined as part of the platform enables a consistency across separate design teams. The environment can be divided into 4 main parts as shown in Figure 6-3:
- Simulation Enablement
- Verification Suite
- Product Realization
- Documentation.

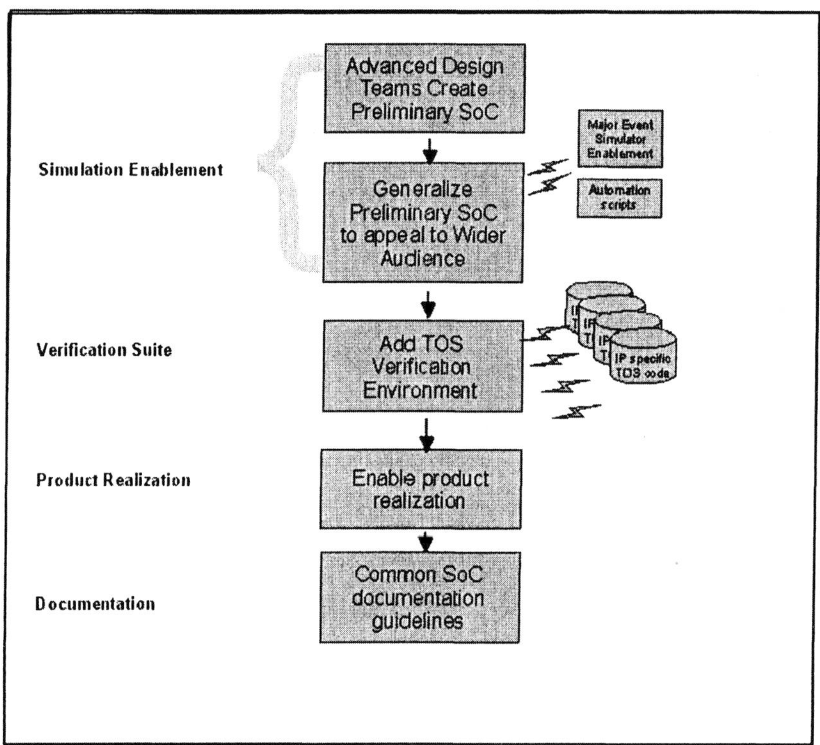

Figure 6-3. Platform Based Design Creation Flow

All of these turn a template or base design into a final product that enables a "Platform Based Design" methodology.

Simulation enablement is a major requirement for the productization. It enables design teams that prefer different event simulators. The idiosyncrasies of the major event simulators have been flushed out and the platform is verified to work with all of them. Simulation enablement also allows more utilization of available platforms/licensing schemes.

The verification suite needs to be flexible, extensive, and portable. The TOS methodology meets or exceeds these requirements.

Product realization for a reference platform should be taken as close to hardware as feasible. IBM has a standard ASIC methodology for product realization that meets the needs of SoC designs. Synthesis, timing, and floorplanning are some examples of efforts that need to be done to bring an SoC to realization. IBM's platform offers examples of synthesis and timing using IBM's standard ASIC methodology. It should be noted that any changes to the SoC would cause negating changes to any efforts in this area. These are offered as examples only to ensure correctness of the platform design in the ASIC process.

Documentation is the most crucial, yet least attractive and rewarding part of any effort. IBM's platform documentation serves two purposes. First, documentation serves as a reference guide for installation and use of the platform itself. Secondly, it can be used as a starting point for many derivative designs. Address maps, descriptions of clocking and initialization are examples of the data that can be taken almost verbatim from the platform's documentation.

IBM not only has attempted to make a methodology for Platform Based Design, but also a methodology for Platform Creation. A bus centric platform must be updated as new buses are developed, but the rules and methods for creating new platforms require consistency.

4.2   Case Study

The concepts above were applied to two low-end network processors developed as ASSPs, the NPe405L and NPe405H network processors. Figure 6-4 depicts the base PowerNP™ block diagram. The NPe405L is a 1.5 million-gate design in 0.25um technology based on the reference platform with an additional Ethernet macro and a 32-channel High Level Data Link Control (HDLC) controller. The NPe405H incorporates two additional Ethernet macros for a total of four, as well as a PCI-to-PLB bridge and an 8-port HDLC controller. At the time the designs were started, all of the cores existed in the Blue Logic library along with the accompanying TOS code except for the HDLC controllers and a new logic block designed to interface with the Ethernet macros. The core development team in Europe responsible for the HDLC macros was trained on TOS and delivered the necessary code to be incorporated into the chip verification environment along with the core logic.

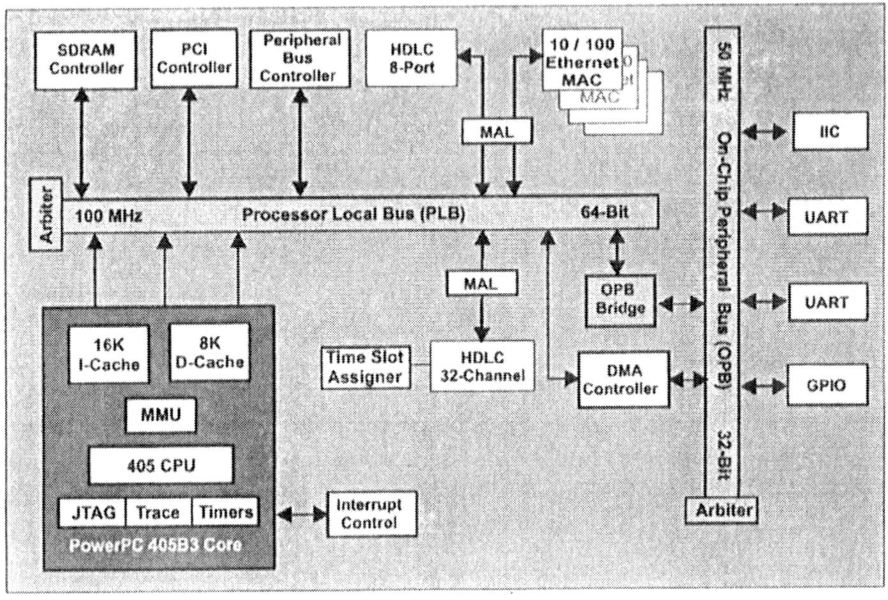

Figure 6-4. PowerNP Block Diagram

The use of the reference platform as a starting point enabled these two similar chips to be developed on overlapping schedules by one design team. One engineer was responsible for the integration work and also was able to spend time working on a customer chip program in parallel since the incremental work of adding a few new cores to an existing platform did not occupy him fulltime. The verification team consisted of two to four engineers at any given time, shared between the two chip programs. One timing engineer was able to work on both programs since they went through physical design at different times. The development of chip-level timing assertions for the first chip allowed the assertions for the second chip to be completed in a much shorter timeframe. Overall, the resource sharing between these programs and the existence of a reference platform to base the designs on allowed the development to proceed very efficiently. Both designs were completed in a twelve-month period with significant parallel development and resource sharing.

As the methodology matured it was extended for use across IBM's WWDC sites around the US and in Japan. This required modifications to allow the platform and tools to work in multiple environments supporting different file systems and various logic simulators. Location-specific information such as hard-coded filename paths needed to be stripped out and the approach made more generic in order for it to work seamlessly under a variety of conditions. The payoff was a portable reference design and methodology that is now in use at sites around the world. Improvements such as the addition of coverage analysis by users at one site, have been incorporated into the methodology, providing a means for user-driven continuous improvement.

Using the platform as a starting point and with the reusable verification methodology based on TOS, IBM was able to reduce chip development cycles significantly for CSSP designs in the communications, storage and printer market segments at sites around the world. Prior to the development of TOS, early CSSP designs (500K gate designs in 0.5um technology) required, on average, about 15 months. By leveraging the initial implementation of the methodology including TOS verification, chip development cycles were reduced to 9-12 months for 1-2M gate designs in 0.25um technology.

As IBM moved to .18um technology, further refinement of the methodology has allowed even more dramatic savings in time and resources for SoC development. IBM's first .18um SoC design was further complicated by the addition of 8Mb embedded DRAM. It also was the first chip to utilize the PowerPC 405, CoreConnect buses and peripheral cores in that technology. The core technology remapping had not been completed prior to the project's start so the design was started in .25um technology and mapped to .18um when the cores became available. The design also incorporated 350K gates of customer logic and the customer's verification environment was integrated into TOS.

Development of that first .18um chip required 12 months, including time spent to bring up the system in a co-simulation environment to facilitate customer software development. Immediately following this project, a derivative design leveraged the now-mature reference design in this new technology and was completed in less than five months. Since then, several other programs have gone from initial architecture development to release to manufacturing in fewer than six months. For example, in 2001, the 405EP product, a derivative ASSP, took only four months and two weeks from concept to release to manufacturing.

5.   PRE-SILICON SOFTWARE DEVELOPMENT

To minimize time-to-market, software development must occur in parallel with hardware development. This requires a pre-silicon environment that mimics the hardware, and simulates fast enough to allow software development. Several approaches are typically used that enable early software and system verification relying on logic simulation and emulation techniques.

5.1   Hardware Software Co-simulation

Current work being done leverages high-level models of the IP using RTL or netlist descriptions in a co-simulation or co-emulation environment. Co-simulation typically uses an instruction set simulator (ISS) of the processor connected to a bus functional model of the processor bus. A source-level debugger is often provided to allow the software developer to work in a familiar environment. The hardware and testbench part of the design is simulated in a logic simulator or accelerator. In such a configuration, the time spent running embedded code on the instruction set simulator is negligible compared with the time spent evaluating the hardware model in the logic simulator. Overall model performance, however, can be an issue when running the system at logic simulator speeds. This is especially true when using event-based logic simulation.

5.2   Bonded-out Emulation

One option for early software development is bonded-out logic. Designers have used this approach to perform system verification and early software development, allowing them to have many lines of boot code and device drivers ready when the hardware is delivered.

The bonded-out approach uses an ASSP on a reference board – typically either the 405GP or 440GP. These ASSPs contain an External Bus Controller core, (EBC), attached to the processor local bus (PLB). The EBC allows data transfers between the PLB and external devices attached to the external peripheral bus. Developers then can synthesize the connecting EBC to PLB logic and the remainder of the unique logic in their design into a Field Programmable Gate Array (FPGA) system connected to the 405GP in hardware. The result is a system that mimics the hardware under development. This system can be used as a means of validating the hardware logic as well as a platform for early software development. The development tools run on the hardware and the performance is limited by the EBC to PLB interface or by the speed of the FPGA.

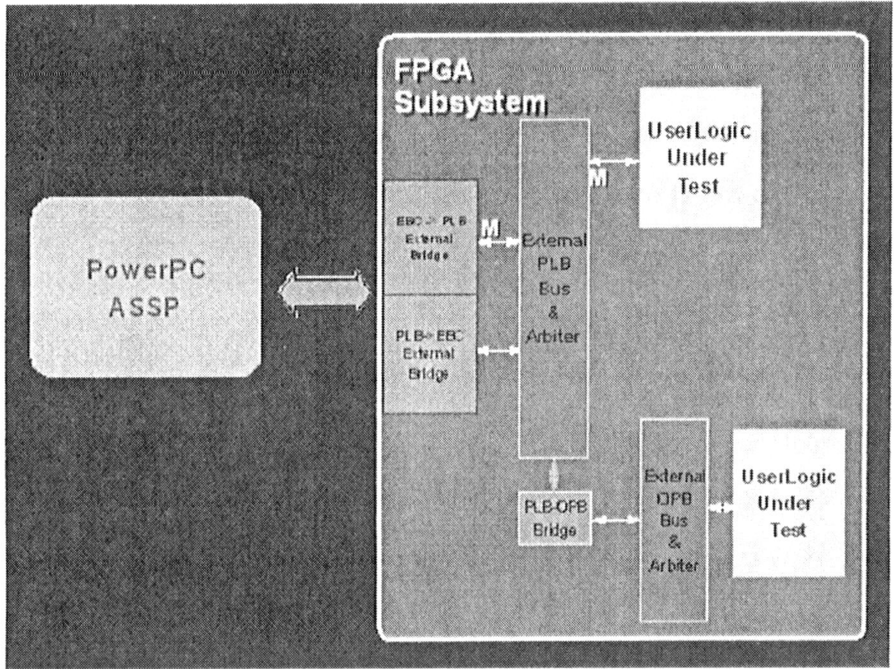

Figure 6-5. FPGA Subsystem

5.3   In-Circuit Simulation (IC-Sim)

IBM created In-Circuit Simulation to help address high performance co-simulation needs. IC-Sim is a transaction-based interface engine between an IBM gate-level accelerator or standard PLI event simulator, and client software programs. The engine delivers software-originated transactions (loads and stores) to bus models in a design, and in turn delivers responses and exceptions back to the software clients. Using the interface, software applications running in an instruction set simulator interact with virtual hardware. When combined with high performance simulation acceleration, meaningful software development and verification can be achieved.

IC-Sim enables software development of pre-silicon hardware due to the visibility of all bus and cache transactions, and enables interaction with HDL, Hardware Verification Language (HVL), or C behavioral languages. This allows software to drive and expect off-chip protocols and traffic essential to software algorithm and driver development and debug. Using the performance advantage of acceleration, along with an ISS, an RTOS can be reasonably booted in the simulation environment. IC-Sim offers an order of magnitude performance advantage over conventional event simulator-based

HW/SW co-simulation systems and a cost advantage over emulation techniques; no extra hardware or RTL partitioning is required. Using IC-Sim, software/hardware data flows can be instrumented to allow measurement of the properties of embedded software - useful for architectural, performance and power analysis based on real life scenarios.

5.4 Using Embedded Software for System Functional Verification

System-level (SoC or ASIC) functional test cases are written according to the specifications of the IP, industry standards and customer input. Such an approach is bottom up, and requires significant development time and validation effort. The approach relies on complete and exhaustive unit-level testing, followed by chip interconnect testing, under the assumption that functional IP, connected properly will produce a functional chip. In reality, all that is proven is that the hardware matches the specification (assuming, generously, that testing is complete and exhaustive), but does not mean the system operates properly for a given customer application.

As SoC and other ASIC designs become more complex, the bottom up verification problem will grow at an exponential rate, and surely will not be appropriate within today's time-to-market and resource pressures. Risk management will become impossible; knowing when you are done will be difficult. A top down, high-level, software-based chip qualification methodology can complement existing verification phases and provide a measure of functional validity.

What is chip functional verification? Ultimately, software applications determine the functionality of the chip. While hardware features and functions specifically used by the application *must* work properly; other hardware functions *may* not need to. This is the essence of verification risk management. It will become cost prohibitive to verify all chip functionality in all its combinations, when only a subset of the combinations will be used by software. How do you know what will be used? What if the application changes in the future and uses unverified paths? The probability of covering these questions during verification is increased by "qualifying" functionality to a level of RTOS and application benchmark standards. Qualification is essentially a stamp of approval that a design meets certain criteria of functionality. It provides a meaningful metric, which is not available today. By combining additional, application specific tests to a qualified base chip design, designers can gain confidence that the final product will function, according to current and future needs.

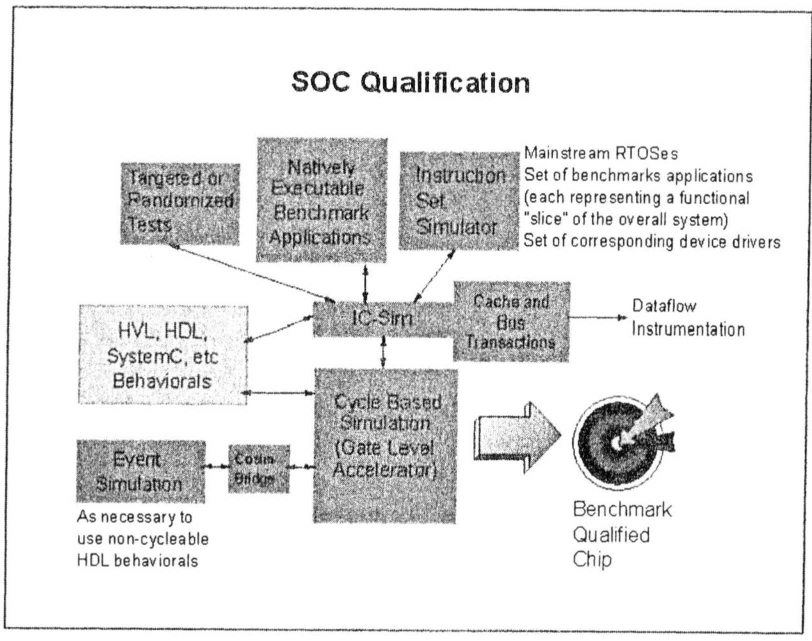

Figure 6-6. Running vertical slices under RTOS control

Chip qualification consists of a combination of a known set of embedded operating systems (RTOSs), standard benchmark application programs, industry standard behaviors/models, and a set of directed/randomized tests. Almost all library IP cores provide some sort of "industry standard" functionality, for example, Ethernet, USB, PCI, and UARTS. The statement of qualification of specific core functionality running in a given chip is obtained from industry benchmark applications. This is known as the vertical slice for the core. Full chip qualification then consists of running a plenitude of simultaneous vertical slices, under control of the RTOS, as shown in Figure 6-6. Combining embedded software development with chip verification methodologies enables a more robust system-wide discipline and will promote a wider scope of reuse than is currently achieved in the SoC field.

6. SUMMARY AND FUTURE DIRECTIONS

A methodology for the development of SoC products must continually focus on the related goals of reducing development time, cost and risk. IBM has addressed these challenges through a strategy of reuse. This reuse is accomplished both in "build-ahead" and "build-as-you-go" forms. The

Platform Based Design, which provides a pre-integrated template of a chip that is staged ready for the next project to continue to a finished product, is an example of "build-ahead" reuse. The verification code for TOS that is developed immediately after the IP and supplied along with it at integration time is an example of "build-as-you-go" reuse. For either form of reuse to be effective, standards must be in place to guide the development so that the content needs little or no modification when it is applied. IBM has implemented standards for interconnection, IP development practices, chip register architectures (for software reuse), and reusable verification code. Each of these standards has provided benefits in one or more areas of SoC development.

Today, in the era of multi-million gate ASICs, reusable designs, verification consumes about 70 percent of the design effort [4]. By segmenting verification into tiers and developing a robust building-block verification environment at the chip level, IBM has leveraged reuse to reduce the overall verification effort.

After an initial effort at producing plug-and-play IP building blocks, IBM found further development leverage through implementation of a Platform Based Design (PBD). This PBD was defined by integrating the most often used IP building blocks from previous designs. Open slots were left on all busses for incorporating customized IP. Included with this "template" are the environment, tools and test code for verification, enablement for realization and documentation to assist with all phases of the development effort. This PBD has enabled significant reductions in labor required for the front-end phases of SoC development and reduced the elapsed time from definition to release from approximately 15 months in 1998 to less than 6 months in many cases.

Providing the ability to develop application software simultaneously with development of the hardware is a further way of shortening the total time-to-market. IBM offers the capability to do this using hardware/software co-simulation and also provides specially developed bonded-out logic for use in third-party emulation equipment. Recently, the capability to achieve many times the throughput of standard co-simulation without the cost of emulation hardware has been developed using IC-Sim. When IC-Sim is used with hardware acceleration, performance adequate to boot an RTOS and debug software can be achieved.

Moving forward, development standards used within IBM will migrate toward use of industry standards currently being developed. IBM participates actively in the groups that are developing these standards such as the VSIA. Use of these standards that were not available when this work began, will provide more options when considering whether to develop or purchase IP, and improve the choice of development tools as EDA companies develop products to these standards. These new industry standards will need to accommodate the challenges of newer, deep sub-micron ASIC technologies, software enablement, and future on-chip architectures for IP block bussing and interconnection.

The future of reusable verification lies not only in reuse across parallel projects, but also from the core to the system level. Currently cores are fully verified prior to check-in to the ASIC library; then, an effort is made to backtrack and create the reusable TOS code for system-level verification. The primary drawback to this is that many of the same test cases may be needed, but in a different environment. This leads to essentially doing the same development and debug work twice. Another problem is that there may be many additional good test cases in the core-level verification suite, that get overlooked. They not only are suitable for system-level tests, but also create unique or interesting scenarios in that realm as well. The solution seems to be the implementation of another reuse methodology and standards, as complete as TOS is today, but in some sort of hardware verification language allowing test case propagation between levels.

The benefits of creating a reusable platform and its entire infrastructure must be clearly shown to justify the significant investment required to produce a robust, validated PBD. It is clear that the more application-specific the PBD, the more downstream effort can be reduced. However, this also means it must be reconfigured more often to incorporate technology improvements and to accommodate customers who choose between different, competing standards. (Consider the choice between Hypertransport, RapidIO and 3GIO.) Substantial work can be done to help automate the generation of the platform itself. Automated integration or "push button" PBD tools emerging now could significantly reduce the cost of accommodating change and competing requirements on the platform itself. These tools can be described as a way of gathering the expert knowledge required to put together a SoC design. It is important to note that the work required to encapsulate and automate that knowledge is not a trivial task. PBD developers also must consider carefully how much control of the platform and access to the source of the platform is given to the end user. The more access, the greater flexibility, but the risk increases that the end product will require substantial work to be verified.

Ultimately the same tools that allow concurrent software and hardware development, can be used to validate that the entire application (hardware and software combined) will perform in the target machine. This system-wide capability when incorporated into an SoC methodology can help reduce cost through eliminating re-spins and shorten development time, but above all can help reduce risk in three areas: hardware, software and the combination of both.

REFERENCES

1. B. Lewis, "ASIC Suppliers Target System-Level Integration," Dataquest, ASIC/SLI Worldwide, Market Trends, April 16, 1998.
2. B. Blaner, D. Czenkusch, R. Devins and S. Stever, "An Embedded PowerPC® SoC for Test and Measurement Applications", *Proceedings of the 13th Annual IEEE International ASIC/SoC Conference*, September 2000.
3. R. Devins, "SOC Verification Software – Test Operating System", *IEEE/DATC Electronic Design Processes Workshop*, April 2001.
4. J. Bergeron, *Writing Testbenches – Functional Verification of HDL Models*, Kluwer, 2000.

The authors would like to acknowledge Scott Brooks and Anne T. Roe for their helpful comments on this manuscript.

Chapter 7

PLATFORM FPGAS
FPGAs for programmable systems

Patrick Lysaght
Xilinx Research Labs

Abstract: Xilinx relies on systematic design reuse to make the benefits of system on chip integration available to the widest design community in an effective and affordable manner. Platform-based design is important in the design of FPGA architectures and is of growing significance in the design of systems incorporating FPGAs. Platform-based design with FPGAs is similar to platform based design with ASICs but noteworthy differences are also evident. The unique features of FPGAs can give rise to new types of platforms, such as the self-reconfiguring platform, that exhibit some completely novel capabilities. Design reuse is become increasingly important in the FPGA community as average design sizes continue to grow. One likely consequence of this trend is that platform-based design will proliferate and evolve further in the hands of the FPGA design community.

Key words: Platform FPGAs, system on chip, design reuse, platform-based design, self-reconfiguring platform

1.   INTRODUCTION

The benefits of increasing levels of integration have been demonstrated continually over several decades in the electronics industry.   With present ASIC technology, it is possible to integrate an entire system on a single chip. This level of integration empowers system designers to produce next-generation products that set new price-performance benchmarks. While we have the capability to produce systems on chip and the market opportunities for them are well established, most design teams do not enjoy access to this kind of state-of-the-art ASIC technology. The engineering efforts and the financial costs associated with producing a system on chip ASIC are so great that only relatively few applications with extremely high volumes can justify the investments required.

At Xilinx, we interpret the challenge of *winning the SOC revolution* as that of making the benefits of SOC technology available to the widest possible design community in an effective and affordable manner. We produce families of Platform FPGAs for high-performance, embedded systems. The Virtex™-II [1] and Virtex™-II Pro [2] devices are optimized for designers who have to meet their customers' requirements within aggressive time-to-market constraints. Our goal is to provide designers with the benefits of leading edge SOC technology while shielding them as much as possible from the complexity of the SOC design process.

Systematic design reuse is central to Xilinx's strategy. On the one hand, we employ extensive reuse and state-of-the-art silicon processes to design some of the most complex custom integrated circuits ever produced. On the other hand, we provide Platforms FPGAs so that the advantages of SoC design are within the reach of any design team. The designers in turn can raise their level of design reuse to accelerate the development of their products.

In this chapter we explore design reuse in the context of Platform FPGAs. We approach the topic from two distinct perspectives, that of the engineer who creates the FPGA and that of the engineer who designs FPGAs into his system. We contrast the deployment of systematic reuse techniques in FPGA design with platform-based design (PBD), the leading design reuse methodology espoused for SoCs. We begin by reviewing platform-based design as it is currently presented and proceed to investigate how well this model applics to Platform FPGAs. In the process we identify a number of significant differences in how PBD is applied to SoC designs in ASICs and in FPGAs.

We posit that, as the performance and capabilities of Platform FPGAs continue to improve, PBD will proliferate among high-end FPGA users. Over time we anticipate that the number of design teams using platform based methodologies for advanced FPGA designs may well exceed their ASIC counterparts by at least one order of magnitude.

Looking to the future, we present an example of a new class of platform that is unique to Xilinx Virtex II FPGAs. It is called a self-reconfiguring platform (SRP) [10] and enables an FPGA to dynamically reconfigure itself via an internal configuration access port (ICAP) under the control of a microprocessor embedded in the array.

2.   **PLATFORM BASED DESIGN**

Platform-based design (PBD) is one of the key strategies for successfully coping with the most complex, system-on-chip designs. Its basic premise is that the levels of design productivity needed to counter the intrinsic complexity of such embedded systems will only be achieved by extensive and planned design reuse [3,4,5,6]. The current status of PBD is summarized by Martin [7] as follows:

> "Platform-based design is the best-validated industrial approach for achieving high reuse in SoC design and incurs the lowest risk in derivative creation ... PBD moves beyond the reuse of individual IP blocks, to the reuse of complex architectures of hardware and software components organized for a specific application."

The use of platform based techniques is also associated with more subtle changes in the nature of the design process. There is greater acceptance of the importance of constraining unnecessary degrees of freedom in the design process. The number of SoC architectures is limited by the enormous cost and risk associated with their development. Increasingly product differentiation is being achieved via programmable components such as embedded microprocessors and FPGAs. Finally, the rise of the consumer electronic sector has placed a premium on getting products to market as quickly as possible. Consequently, it is now more important to design a system that meets the target specifications *on time* than to design a solution with better specifications that delays the introduction of a product to the marketplace.

2.1   **Different platform classes**

The platform-based methodology has evolved significantly beyond its original formulation. In Table 7-1, four classes of platforms are identified and examples of each are presented [8].

Table 7-1. Platform types

Platform type	Example
Full-application platforms	Nexperia, Philips Semiconductors Open Multimedia Applications Platform (OMAP), TI
Processor-centric platforms	ARM Micropack
Communication-centric platforms	Sonics uNetwork
Fully programmable platforms	Xilinx Virtex-II Pro

In the next section we explore the phenomenon of the last of these four categories, fully programmable platforms, using the Xilinx Virtex-II Pro family of devices as our reference model.

3.   VIRTEX-II PRO FPGAS

Xilinx offers two families of Platform FPGAs. The first of these is the Virtex-II family, which is described as a platform for programmable logic. The second is the Virtex-II Pro family, which is described as a platform for programmable systems. The Virtex-II Pro series is a direct derivative of the highly successful Virtex-II family.

3.1   Silicon Process Technology

The architecture of any FPGA is intrinsically programmable. Providing programmability results in smaller functional capacities and lower operating speeds than ASICs fabricated with equivalent technology. However, by migrating to leading edge silicon processes, FPGAs are now large enough and fast enough to be used in the majority of applications in which ASICs are used. In contrast, most ASIC designs will be fabricated in older, less advanced processes because the non-recurring engineering charges (NRE) associated with newer processes are so exorbitant.

Virtex-II Pro Platform FPGAs are designed as custom integrated circuits in 0.13 micron process technology using high-performance 92-nm transistors. The largest devices in the family contain more than three hundred million transistors. They also feature a combination of nine layers of all-copper interconnection, low-k dielectric and a 1.5 V core voltage. Virtex-II Pro devices are produced on the latest 300-mm wafers and exploit flip-chip packaging to achieve package pin counts of up to twelve hundred pins.

The combination of 0.13 micron process geometries and 300-mm wafer production are critical to the competitiveness of Platform FPGAs. Xilinx's crucial contribution is that we design with *bleeding-edge* technologies so that our customers do not have to. In effect, we *tame* the technology so that our customers can enjoy the benefits without enduring huge engineering problems and non-recurring engineering charges. To do this, Xilinx relies heavily on systematic design reuse.

3.2 Systematic Reuse

The Virtex-II and Virtex-II Pro share a common programmable architecture consisting of the following key components:
- Logic fabric
- Routing
- Input/output
- Block SelectRAM
- Multipliers
- Digital clock managers

Since the Virtex-II architecture was implemented in an older 0.15 micron silicon process with eight layers of metal, it was necessary to migrate the architecture to the new 0.13 micron, nine-layer metal process used for the Virtex-II Pro family of parts. To emphasize the extent of the design reuse, the characteristics of common components are briefly described.

3.2.1 Logic fabric

The programmable fabric is arranged as a two-dimensional array of configurable logic blocks (CLB), each of which is composed of four slices and two tri-state buffers. A slice contains two 4-input, look-up tables and two flip-flops. Each LUT can also be configured to operate as a programmable 16-bit shift register or as a 16-bit distributed RAM. Fast arithmetic logic gates and fast look-ahead carry-chains are also provided. The CLB contains all the associated routing for interconnecting its internal resources and connects to a switch matrix to access general routing resources.

3.2.2 Routing

Interspersed between the rows and columns of CLBs is an abundance of programmable routing and switch boxes. The segmented routing structure uses active interconnect technology. This results in fast, predictable routing delay, which is independent of fan-out and provides immunity from deep sub-micron induced crosstalk. The routing hierarchy consists of short, medium and long-range programmable routes and is symmetrical in the horizontal and vertical directions. All the hard IP blocks use the same interconnection scheme and have the same access to the global routing matrix.

3.2.3   Input/Output

The array is surrounded by a periphery of programmable I/Os. A total of twenty-two single-ended standards and five differential standards are supported. The largest device has approximately 1,200 pins. The maximum I/O speed is 840 Mbps which is available in the low voltage differential signaling (LVDS) mode. A feature called digitally controlled impedance provides on-chip termination for single-ended I/O standards.

3.2.4   Block SelectRAM

The architectures feature a large number of block SelectRAMs (BRAMs) each of which is an 18kbit dual-port RAM. The BRAMs are programmable from 16kx1 to 512x36 in various width and depth combinations. The ports are synchronous and BRAMs can be cascaded to implement larger embedded storage blocks.

3.2.5   Multipliers

There is a dedicated 18 bit x 18 bit signed multiplier (2's complement) block associated with every BRAM. These multipliers are optimized for operations using the BRAM content as one operand, although they can be used independently.

3.2.6   Digital Clock Managers

Multiple digital clock managers (DCMs) are provided for clock de-skew, frequency synthesis and phase shifting. The DCMs provide multiple system clocks that are phase aligned with a system reference thus eliminating clock distribution delays. Frequency synthesis permits accurate clock multiplication or division. Finally, both coarse and fine-grained phase shifts are supported with dynamic control to compensate for temperature and voltage variations.

3.3   New hard IP Core Functionality

The Virtex-II Pro series is distinguished by the inclusion of two new kinds of hard IP blocks. The first of these is the PowerPC 405[TM] from IBM. The second is the RocketIO[TM] serial transceiver based on Mindspeed's SkyRail[TM] technology.

3.3.1   PowerPC 405 RISC Core

The PowerPC 405 RISC core is a 32-bit, Harvard style CPU that is capable of operating at 300 MHz and 420 Dhrystone MIPS. Its main features include:
- Five stage datapath pipeline
- Hardware multiply and divide unit
- Thirty-two 32-bit general purpose registers
- Separate 16 kbyte, two-way, set-associative instruction and data caches
- Embedded memory management unit with translation look-aside buffers and variable pages sizes
- Dedicated on-chip memory interfaces
- Timer facilities
- Debug and trace support

The integration of a high-performance microprocessor into the fabric of the FPGA requires careful design. The goal is to obtain the best advantages of the microprocessor and the FPGA by creating a tight coupling between them so that the overall system performance is optimized. Specially designed interface logic integrates the core with the surrounding CLBs, block RAMs and general routing resource.

The PowerPC 405 uses IBM's CoreConnect^{TM} bus hierarchy, which comprises three separate buses, for interconnecting processor blocks, IP blocks and custom logic. The three buses are the processor local bus (PLB), the on-chip peripheral bus (OPB) and the device control register (DCR) bus. High-performance peripherals connect to the high bandwidth, low-latency PLB. Slower peripheral cores connect to the OPB, thus reducing traffic on the PLB and improving overall system performance. The separate DCR bus is used to manage the status and control registers of peripherals external to the PPC 405. It has its own unique address space which improves both system integrity and performance. To maximize flexibility, the CoreConnect buses are offered as soft IP so that each designer can synthesize the bus combination that best suits his particular system requirements.

Fig 7-1 is a die photograph of the XCVP7 Virtex-II Pro FPGA (after removal of the nine layers of metal). The XCVP7 has one PowerPC and eight RocketIO^{TM} multi-gigabit transceivers. The PowerPC 405 is clearly visible in the center of the photograph while the smaller transceiver blocks are arranged in groups of four along the top and bottom edges of the die. The darker vertical columns are the BRAMs. Note in particular how well the PowerPC is integrated with the internal memory structure of the FPGA.

Xilinx invented a patented technique called IP immersion to ease the process of embedding hard cores such as the PowerPC into the FPGA fabric.

The PowerPC core uses only the first four layers of metal interconnect so that the remaining layers are preserved for the programmable routing. The fact that embedding of the PowerPC in the logic fabric does not interrupt the programmable routing structure greatly simplifies the task of the automatic routing software during the placement and routing stage.

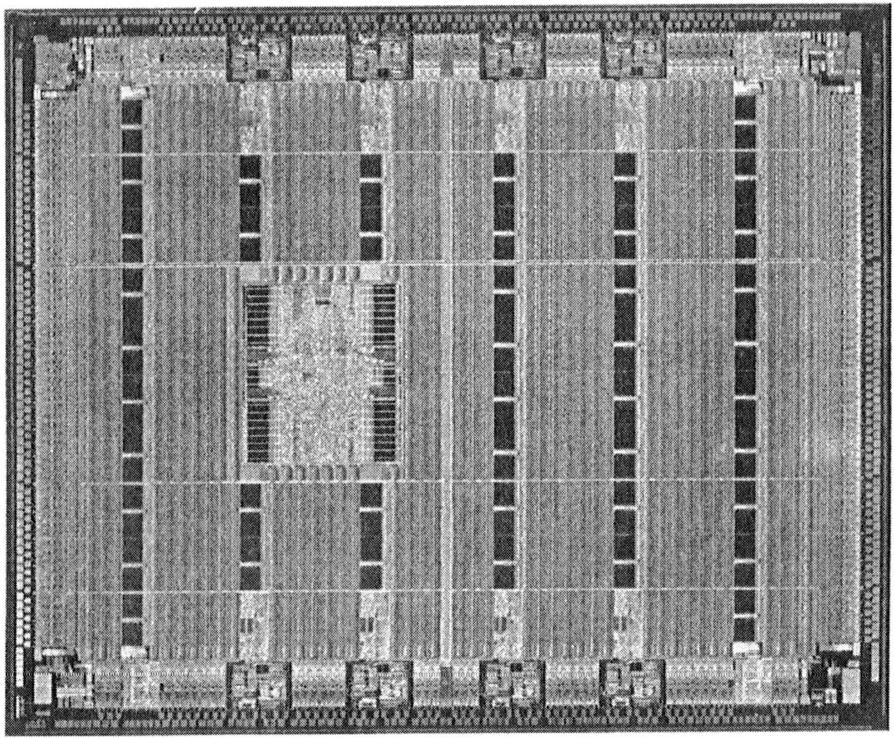

Figure 7-1. Die photograph of the XC2VP7 VirtexII Pro FPGA

3.3.2   Multi-gigabit Serial Transceivers

The second hard core is the RocketIO^{TM} serial transceiver which is capable of operating at data rates up to 3.125 Gbps. Starting with the smallest devices, four transceivers are available while the largest devices can support 24 transceivers. The key features of the multi-gigabit transceivers are as follows:
- Full-duplex serial transceiver (SERDES) operating from 622 Mbps to 3.125 Gbps
- Monolithic clock and data recovery (CDR)

- Fibre Channel, Gigabit Ethernet, 10 Gbps Attachment Unit Interface (XAUI), and Infinband-compliant transceivers
- 8-, 16-, 32-bit selectable internal interface to FPGA
- 8B/10B encoder and decoder
- 50/70 Ohm programmable on-chip transmit and receive terminations
- Programmable comma detection
- Channel bonding support for two to sixteen channels with elastic buffers for inter-chip de-skewing and channel-to-channel alignment
- Rate matching via insertion/deletion characters
- Programmable pre-emphasis and output differential voltage levels

Virtex-II Pros are available with 1, 2 or 4 PowerPC 405 cores immersed in the fabric of the FPGA. Starting with the smallest devices, four RocketIO™ transceivers are available while the largest devices can support 24 transceivers.

3.4   Reuse

Clearly, systematic design reuse played a central role in the development of the Virtex-II Pro FPGAs. While "the reuse of complex architectures of hardware and software components" is clearly achieved it could be argued that neither of the Virtex-II architectures is "organized for a specific application". Although the Xilinx Virtex-II Pro series is a direct derivative of the Xilinx Virtex-II, both the original platform and its derivatives are platforms in their own right. To be fair, the success of FPGAs in the high-end communications market has meant that Virtex-II devices were designed to be high-performance parts that are especially compatible with the needs of the communications market. Nonetheless, they remain generic platforms that are used in a range of different applications. It is clear that a platform is being reused but in a manner that is peculiar to the development of new FPGAs.

3.4.1   Partnering for success

One characteristic shared by companies that are successful in raising their levels of design reuse is the extent to which they engage in partnerships to achieve this goal. Their willingness to collaborate with third-party companies reflects their confidence in their core competency and the value proposition that they offer. This gives them the base from which to partner with third parties whose expertise complements their own. Without the reuse of third-party IP, the level of design reuse that can be achieved is typically much lower.

The full extent of these partnerships is not always apparent at first glance. In the case of Xilinx such partnerships are at the heart of our business model. In the first instance, Xilinx pioneered the *fabless* semiconductor company model so that we use partners such as IBM and UMC to fabricate our FPGAs. As the Virtex-II Pro demonstrates, we have also licensed hard IP cores such as the PowerPC and soft IP in the form of the synthesizable CoreConnect bus hierarchy. We also license embedded FPGA core technology to IBM so that we are both a provider and consumer of hard IP. Both Virtex-II families share an extensive library of soft IP cores. Xilinx develops the majority of these cores but third-party vendors contribute many others.

3.4.2   Software Reuse

Clearly having the best FPGA architectures is critically important but on its own it is not enough. Xilinx spends as much engineering effort as the rest of the programmable logic industry combined to ensure that the quality of its software tools matches that of its Platform FPGAs. The two Virtex-II families share a common set of CAD tools for synthesis, mapping, placement and routing, embedded software development and bitstream generation. In addition to substantial in-house development of CAD tools, Xilinx collaborates extensively with leading CAD companies in the development of new and improved CAD tools, especially for synthesis and simulation.

Xilinx's Integrated Software Environment (ISE) is the leading integrated design environment for programmable system on chip design. It consists of design tools for logic design, embedded design and system design. It combines three key qualities that make it unique:
–   Outstanding quality of results
–   Exceptionally fast run times
–   Very low cost
The investment in CAD tools that has been leveraged across Xilinx FPGA families means that the SOC designer working with a Platform FPGA can meet his timing closure requirements quickly, so that he can complete design iterations more effectively and enjoy greater productivity. When combined with the low cost of ownership, these are compelling advantages, especially now that the design costs of producing an ASIC SOC are becoming as exorbitant as the NRE costs. *Winning the SOC revolution* requires making the benefits of SOC technology available to the widest possible design community in an effective and affordable manner. Through the combination of its Virtex-II Platform FPGAs and ISE development tools, Xilinx is achieving this goal.

4.   PBD WITH VIRTEX-II PRO FPGAS

Consider a typical scenario in which a provider of Internet-enabled embedded systems may have a range of products that he chooses to implement on Virtex-II Pro devices. It makes sense that he will ensure that all of these products share common hardware and software components. In this case he may choose to use the PowerPC running Linux as his system processor and the CoreConnect bus hierarchy for his peripherals. Next he selects a high-speed Ethernet MAC and a software protocol stack, such as TCP/IP, for his network. The result is a user-specific system platform that has been defined by the user without direct vendor involvement.

We see the platform definition stage as essentially a bottom-up design process. In completing the design the user will complement this approach with a top-down exploration of the remaining design space. He must add software cores, user-defined logic, and system and application software as necessary. It is during this phase that he seeks to differentiate himself from his competition. Finally, he must integrate the platform with the other components of the system, perhaps adapting his original platform in the process. The end result is a design derived from his unique platform. For subsequent designs, he reuses the platform thereby amortizing the cost of its development over several designs.

With the use-model we have projected, the user, *not* the silicon vendor, creates the platform for a given application domain. The final designs are derived from the user platform.

The designer using Platform FPGAs does not have the same degrees of freedom that an ASIC SoC designer experiences. For example, in the case of the Virtex-II Pro user, choosing between parts in the family is the only control he can exercise over the hard IP present in his chip. It is clear that the Platform FPGA cannot optimally meet every designer's individual specifications. But so long as the FPGA architecture is well-engineered and the family members provide adequate granularity of resource choice, the programmability of the solution more than compensates. Moreover the cost and time-to-market advantages of not having to fabricate a dedicated ASIC far outweigh any disadvantages. In the final analysis, the overwhelming success of the Virtex-II devices is clearly the best vindication of the architecture.

Figure 7-2 depicts the three main elements of the design flow. The bottom-up component entails the systematic reuse of legacy infrastructure and corresponds to the platform based design. The top-down component centers on the introduction of new functionality and the determination of a suitable architectural mapping for the new elements. The final stage comprises the integration of the platform and the new functionality.

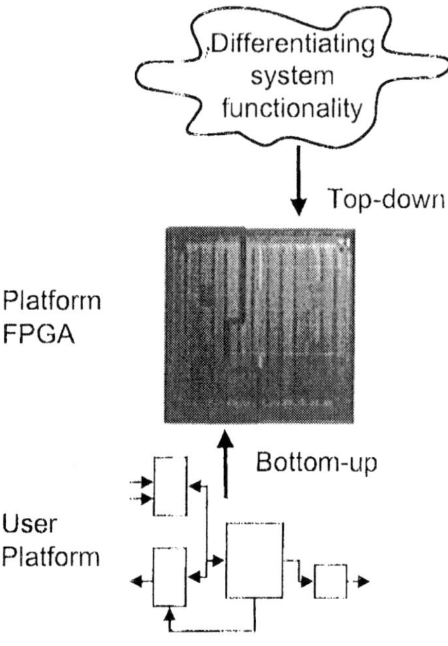

Figure 7-2. Bottom-up and top-down flows

In the process of integrating new functionality into the existing infrastructure, a new, more powerful platform may emerge. Ideally, this in turn will become the platform for the next-generation of projects.

The experience of engineers familiar with PBD in ASIC SoC design suggests a common source of frustration that they have encountered in trying to apply PBD techniques with ASICs: it is very difficult to restrain the natural tendency of engineers to improve the platform. Their experience was that the platform was frequently changed for marginal improvements, which triggered lengthy re-verification. They perceived an advantage of the Platform FPGA was that it eliminated this tendency as far as hard IP blocks were concerned and upheld PBD principles more effectively early in the design cycle.

4.1   Future Platform FPGAs

One perennially safe prediction is that future Platform FPGAs will be larger, faster and encompass greater functionality. Attempts at quantifying the size of these devices by extrapolation and reference to the ITRS roadmap [12] would suggest that by 2005, we could see devices produced with 50 to 70 nanometer technology with the following specifications:

- 1 billion transistors
- 40 million system gates
- 2000 input/output (I/O) pins
- I/O rates of 1.5 Gbps
- Array clock rates of 750 MHz
- Hard CPU speeds of 1 GHz
- Core voltage of 1.0 V
- 24 MB of internal memory
- 165 million configuration bits

Building a one billion transistor platform is an awesome prospect. The architecture of such a device is not just a simple extrapolation of today's architectures. A balance must be found between the processing, storage and communications capabilities that are coherent for each part across a family of devices.

It is likely that the trend toward greater system heterogeneity will also continue in the future. This of course makes the architectural balance even harder to achieve. More powerful and sophisticated microprocessors and digital signal processing blocks are likely to be included in future platforms. Typically this would extend to wider and faster datapath units and more sophisticated cache and memory management arrangements. It may also become viable to integrate non-volatile storage although it may make more sense to fabricate these as separate dies and package them together.

Among analog blocks, analog-to-digital and digital-to-analog converters are most often suggested for inclusion. Even more interesting might be the integration of a complete, high-speed Internet interface. Its inclusion could be justified by using it as a multi-function interface. For example, during design it could be used as a high-speed debug interface, which would justify its inclusion in a general-purpose part.

What we can say with certainty is that the creators of these FPGAs and the designers who design them into future systems will need to achieve the highest possible levels of design reuse. Failure to do so will mean lengthy design cycles and missed market opportunities. Barring new research breakthroughs in system-level synthesis, there is no practical alternative to platform based design as the principal method of coping with design complexity.

5. SELF-RECONFIGURING PLATFORMS

In this section we describe an example of design reuse based on a totally new kind of platform that is unique to Xilinx FPGAs. The self-reconfiguring platform has recently been developed within Xilinx Research Labs and is intended for use in advanced applications of dynamic reconfiguration. For example, a 928x928 crossbar switch for a telecommunication switching application was recently developed on a Xilinx XC2V6000 FPGA [10]. Dynamic reconfiguration of the FPGA's programmable interconnect resources was required to realize the design as it would otherwise have been too big to fit on even the largest Xilinx FPGAs. The self-reconfiguring platform allows us to implement the switch control algorithm in software on the PowerPC and to integrate the entire system into a Virtex-II Pro single device.

5.1   Self-reconfiguration

Dynamic reconfiguration and self-reconfiguration are two of the more advanced forms of FPGA reconfigurability. Dynamic reconfiguration implies that an active array may be partially reconfigured, while ensuring the correct operation of those active circuits that are not being changed. Self-reconfiguration extends the concept of dynamic reconfigurability. It assumes that specific circuits on the logic array are used to control the reconfiguration of other parts of the FPGA. Clearly the integrity of the control circuits must be guaranteed during reconfiguration, so by definition self-control is a specialized form of dynamic reconfiguration.

Currently both dynamic reconfiguration and self-reconfiguration rely on an external reconfiguration control interface to boot an FPGA when power is first applied or the device is reset. Once initially configured, self-control requires an internal reconfiguration interface that can be driven by the logic configured on the logic array.

5.2   Platform Hardware Architecture

The hardware component of the self-reconfiguring platform (*SRP*) is composed of the internal configuration access port (ICAP), some control logic, a small configuration cache, and an embedded processor. These peripherals communicate over the CoreConnect Open Peripheral Bus (OPB). Figure 7-3 shows a block diagram of the hardware subsystem. In this implementation a 32 bit register is used to interface to the ICAP port. Figure 7-3 shows how this register is mapped to the ICAP signals. The control logic for reading and writing data to the ICAP is implemented in a low-level software driver. This driver defines methods for reading and writing to the ICAP interface register. There are also methods for reading and writing blocks of data to the ICAP. The BRAM shown in Figure 7-3 is used to cache configuration data. To keep the *SRP* hardware as lightweight as possible only one BRAM is used. One Virtex-II BRAM can store 18K bits of data. The fundamental unit of configuration in Virtex-II devices is a called a configuration *frame*. A configuration frame is one-bit wide. The largest Virtex-II Pro device, the XC2VP125, has a frame length of 11K bits. So one BRAM can easily store a whole frame, of even this largest Virtex-II Pro device. We have implemented this hardware system on both Virtex-II and Virtex-II Pro devices. On the Virtex-II device we used the MicroBlaze soft processor and on Virtex-II Pro we targeted the embedded PowerPC.

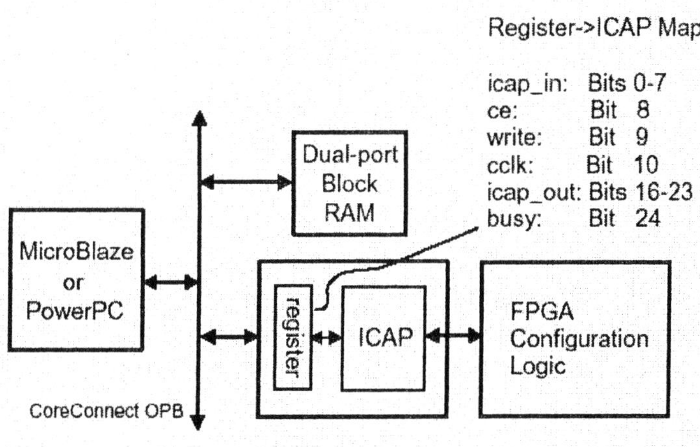

Figure 7-3. SRP Software Architecture

The embedded processor provides intelligent control of device reconfiguration at run-time. The integration of this functionality is especially attractive for embedded systems. This lightweight approach maximizes flexibility while minimizing additional external circuitry.

5.3 Platform Software Architecture

The software component of the SRP is designed as a software stack. Figure 7-4 shows the components of this stack. The software stack is divided into hardware dependent and hardware independent parts. The ICAP API enables this division. This API defines methods for transferring data between the configuration cache implemented in BRAM and the active configuration memory. It also provides methods for accessing the configuration cache. We have created two implementations of the ICAP API. The first uses the SRP hardware described in the previous section to enable embedded applications to read back or modify the active configuration of the FPGA. The second implementation emulates the active configuration store and the configuration cache entirely in software. This emulated version of the ICAP API can then be compiled for a Windows or Unix workstation.

The Xilinx Partial Reconfiguration Toolkit (XPART) is built on top of the ICAP API. Thus XPART does not rely on the underlying hardware and can be compiled for an embedded system or for a Windows or Unix system. This portability also applies to applications that use XPART and/or the ICAP API.

One advantage of this portability is that one can do a degree of debug on a standard workstation before moving to the target embedded platform. It is also possible to use XPART on a workstation to generate static partial bitstreams.

The purpose of the XPART layer is to allow embedded processors to modify resources and relocate modules. XPART provides functions for on-the-fly resource modification. This is done via *getCLBBits()* and *setCLBBits()* methods. These methods abstract away the bitstream and provide seemingly random access to selected resources. Currently XPART has the capability to read or modify all Virtex II CLB logic and routing resources.

Looking to the future, we are excited by the prospect of the SRP being applied in a range of innovative applications that would previously not have been possible without this type of platform.

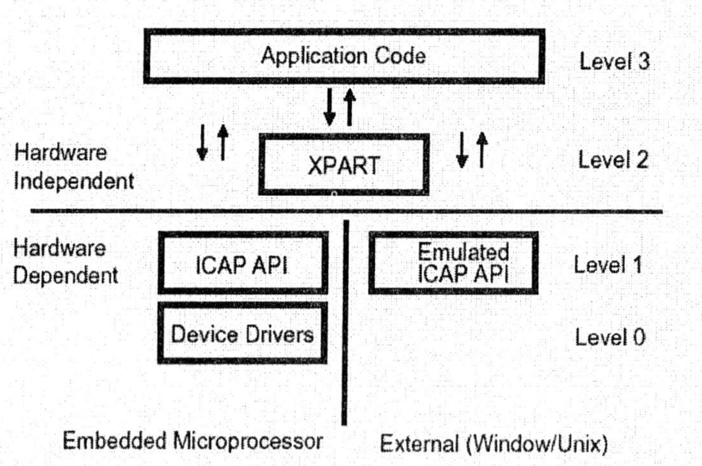

Figure 7-4. SRP Software Architecture

6.   CONCLUSIONS

With Virtex-II and Virtex-II Pro Platform FPGAs, Xilinx makes a key contribution to *winning the SoC revolution* by making the benefits of SoC technology available to virtually any design team. The development of the Virtex-II Pro family from the Virtex-II is a classic example of systematic design reuse, applied in this case to the development of one of the world's most complex custom integrated circuits. The reuse of a common architecture between the two Virtex-II families was complemented by the reuse of critical new hard IP blocks in the form of the PowerPC 405 and RocketIO serial transceiver from external partners.

We demonstrated the viability of PBD methodologies for designers using FPGAs with reference to a typical design scenario. Given the projected increase in device capacity and complexity, we believe that PBD will become increasing relevant to designers exploiting Platform FPGAs in their systems. Perhaps the most significant contribution of Virtex-II Pro platforms will be the proliferation of the platform design methodology beyond the small community of high-end, ASIC designers. It requires a shift of this magnitude to accelerate the evolution of the methodology.

Finally, we presented a novel type of platform that incorporates unique features of FPGAs to allow designers to build self-reconfiguring applications.

ACKNOWLEDGEMENTS

I am grateful to my many colleagues in Xilinx whose work and ideas are reflected in this chapter. I would like to especially acknowledge the encouragement provided by Ivo Bolsens, and the patience of Grant Martin as deadlines expired and new ones replaced them.

REFERENCES

1.   Virtex™-II Platform FPGA Handbook (v1.3), Xilinx, December, 2001.
2.   Virtex™-II Pro Platform FPGA Handbook (v1.0), Xilinx, January 2002.
3.   A. Ferrari and A. Sangiovanni-Vincentelli, "System Design: Traditional Concepts and new Paradigms", *International Conference on Computer Design*, IEEE CS Press, Los Alamitos, Ca., 1999, pp. 2-12.
4.   H. Chang, et al, *Surviving the SOC Revolution, A Guide to Platform-based Design*, Kluwer Academic Publishers, USA, 1999, pp. 1-10.

5. K. Keutzer et al, "System-Level Design: Orthogonalization of Concerns and Platform-Based Design", *Transactions on Computer-Aided Design of Integrated Circuits and Systems*, IEEE, Volume 19, Issue 12, Dec 2000, pp. 1523-1543.
6. A. Sangiovanni-Vincentelli and G. Martin, "Platform-Based Design and Software Design Methodology for Embedded Systems", *Design & Test of Computers*, IEEE, Nov. – Dec. 2001, pp. 23-33.
7. G. Martin, *The Reuse of Complex Architectures,* Guest Editor's Introduction, IEEE Design & Test of Computers, Nov-Dec 2002.
8. G. Martin and F. Schirrmeister, "A Design Chain for Embedded Systems", *Computer*, IEEE, Volume 35, Issue 3, March 2002, pp. 100-103.
9. B. Blodget, S. McMillan and P. Lysaght, "A lightweight approach for embedded reconfiguration of FPGAs", *DATE*, Munich, March 2003.
10. S. Young, P. Alfke, C. Fewer, S. McMillan, B. Blodget and D. Levi, "A High I/O Reconfigurable Crossbar Switch", *2003 IEEE Symposium on Field-Programmable Custom Computing Machines* (to appear).
11. The International Roadmap for Semiconductors, http://public.itrs.net, 2001.

Chapter 8

SOPC BUILDER: PERFORMANCE BY DESIGN

Jesse Kempa, Sheac Y. Lim, Chris Robinson, Joel A. Seely
Altera Corporation[1]

Abstract: We are now in the era of programmable logic. The need for low-cost, high-performance methods of getting to market in the shortest amount of time are ever present. This requires tools and methodologies that leverage the strengths of the designers, utilize known-to-work components, and allow rapid experimentation and innovation. Altera's SOPC Builder is this tool. SOPC Builder is used to construct embedded "Systems on a Programmable Chip" (SOPC) from a wide variety of components that are available natively (the Nios soft-core microprocessor, communications IP, memory interfaces, timers, etc.) or from 3rd party IP providers. Additionally it is a framework into which developers can place their own custom IP cores for easy reuse. The power of this tool and design methodology lies in the fact that it abstracts away the tedious tasks, such as developing the bus wiring, arbitration logic, and memory map decode, and allows the designer to spend more time on architectural issues, such as system performance or software/hardware co-design. With SOPC Builder the designer is able to iterate through designs quickly, testing which one will provide the most optimal performance for the given problem, testing things like simultaneous multi-master architectures, multi-processors, and where to make the split between software algorithms and hardware acceleration. Working systems are generated in minutes, and architectural evolution can begin immediately, giving faster time-to-market and providing a perfectly tailored solution in the end.

Key words: Altera, Nios, SOPC Builder, FPGA, PLD, Programmable Logic, Avalon, Microprocessor, Microcontroller, System on a Programmable Chip

---

[1] *Altera is a trademark and service mark of Altera Corporation in the United States and other countries. Altera products are the intellectual property of Altera Corporation and are protected by copyright and other intellectual property laws and one or more U.S. and foreign patents and patent applications.*

1. INTRODUCTION TO PROGRAMMABLE LOGIC

Programmable logic meets a diverse set of needs. It is an ideal platform for rapid product development because it provides the flexibility of rapid design evolution. Additionally in areas of product uncertainty it provides a low-risk alternative to ASICs. Furthermore the conversion path from the FPGA design into an ASIC can be staged starting from a large device for development to a smaller device for preproduction; it is then possible to take the design and create a HardCopy™ design or gate-array, and finally to move to ASIC if the size, power, and cost constraints require this final step.

Whether the design eventually resides in an ASIC or stays in an FPGA for its entire life is immaterial to the fact that rapid prototyping is essential in today's marketplace. FPGAs are ideal for experimenting and iterating designs. However, once iteration of designs is a realistic option for design engineers, their attention turns to architectural system performance enhancements. This requires the next level of tool integration to allow a system designer the flexibility of trying different architectures that flow between hardware and software implementation seamlessly.

The early pioneers of programmable logic devices were guided by the vision of making hardware design and development as easy and flexible as software development. To achieve this vision many milestones had to be reached. Some of the important milestones include developing programmable logic chips that take advantage of each new process technology, researching the most efficient means of implementing a design written in an HDL such as Verilog or VHDL into a programmable logic chip, perfecting the tools that turn HDL into physical placement automatically, creating or partnering with third-party IP suppliers to create a portfolio of reusable parameterized IP, and developing a tool that can be used as an architectural framework for the use of this IP. To provide software-like hardware development, completing all these steps is necessary.

And so we have done this. We are at an exciting point in time right now, which is the "Era of Programmable Logic".

2. EMBEDDED SYSTEMS ON A PROGRAMMABLE CHIP

Microcontrollers are the original "System on Chip" (SoC) devices. The elements required to create viable custom microcontrollers include: full-featured 32-bit RISC microprocessors, readily available peripherals as well as options for building custom peripherals, intellectual property, and comprehensive development tools. In combination with low-cost FPGAs, designers can use these elements to realize attractive alternatives to off-the-shelf microcontrollers.

The general purpose single-chip microprocessor has found uses in everything imaginable in the past 30+ years. It has also enjoyed huge advances in capabilities and raw compute power. In its evolution, increasingly complex instructions gave way to an analysis of how compilers used these instructions and the debate between CISC and RISC began. In compute-intensive applications the continued evolution for CISC microprocessors moved into the realm of the DSP. In some applications, however, a subset of DSP-like functionality is required, in which case designers have to decide between options such as coding in assembly language and using a higher-clock frequency processor, or using a DSP processor as the main processor, or using a co-processor for the compute intensive portion of the problem.

With embedded microcontroller systems in programmable logic, another set of options are available. The designer is now free to take specific software blocks and code these as hardware that can be called as single "custom instructions" from the microprocessor. From the software engineers perspective this is simply a function call in C or assembly language. It doesn't matter to them that rather than a sequential list of instructions being called, a block of hardware is used to execute the relevant algorithm. Similarly, the architecture of the system can be changed to match the problem at hand, including adding multiple arbitrated slave-peripherals on a multi-master bus, or including custom peripherals or DMA as necessary.

3.   SOPC BUILDER

Systems on a Programmable Chip can be created by hand by piecing together pieces of Intellectual Property (IP), whether developed internally, purchased from a third-party, or found on the web. However the tedium of wiring these components together and the possibility of making errors limit the size and complexity of projects engineers are willing to undertake – especially if they have to integrate untested IP from a third party. This pain indicates an opportunity to create a tool that can take various pieces of IP that conform to a non-proprietary interface specification and wire them together into one complete system. Altera's SOPC Builder is an example of such a tool.

SOPC Builder is an automated system development tool that dramatically simplifies the task of assembling components into complete systems-on-a-programmable-chip. Originally the systems created by SOPC Builder were microprocessor-centric including processors, buses, memories, co-processors, and external interfaces. However, SOPC Builder is more general than this and can be used to wire up any number of embedded systems on a programmable chip [1].

The power of SOPC Builder is that it provides a repository of tested IP components and wires them up for you. SOPC Builder will automatically generate address decode logic and interrupt controllers (IRQs), based on the peripheral memory map and interrupt priorities specified by the user within the user interface (UI) [2]. If a system contains multiple masters (such as two processors, or a processor and a DMA peripheral), SOPC Builder automatically generates the arbitration logic to intelligently connect them. It uses slave-side arbitration to optimize system performance by enabling simultaneous multi-mastering on the Avalon bus. In addition to hardware-generation, SOPC Builder automatically generates header files, peripheral drivers and custom software libraries specific to the system under design.

3.1   Available Components

SOPC Builder provides an extensive list of available components (see Table 8-1). A reasonably complete set of usable IP is a necessary requirement for an architectural integration tool to be effective. However, no tool can provide all things to all people, which is why having the ability to add to the list for the particular application at hand is critical for the power of this tool to be realized. With SOPC Builder this can be accomplished through the use of custom instructions and custom peripherals. Creating and adding components to SOPC Builder is easy and makes them available for all future projects.

Table 8-1. Partial List of Components available through SOPC Builder

Interface to User Logic	Communication
Microcontrollers	USB 1.1 and 2.0
Nios 16 and 32-bit	SPI
1655X	UART (RS-232)
6811	CAN
ARM 922T	I2C
Bridges and Interfaces	802.11
AHB to Avalon	Single and Multi-Channel HDLC
Avalon to TriState	**DSP Functions**
AHB to PCI	DCT
Avalon/PCI	Data Encoder/Decoder
PowerPC Interface	CCIR-656 Encoder/Decoder

Memory	JPEG Encoder/Decoder
Interface to On Chip RAM/ROM	MPEG2/MPEG4
Asynchronous SRAM	**Other**
SSRAM	Instruction and Data Cache
SDRAM	Parallel I/O
Flash	Timer
Serial Memory Interface	PWM
DMA	IDE
Ethernet	VGA Controller
Interface to 10/100 MAC/PHY	Keyboard/Mouse Controller
10/100/1000 On-Chip MACs	Color Space Converter

4.   PERFORMANCE OPTIMIZATIONS

Once over the hurdle of providing a full suite of tools for hardware and software engineers, IP, and peripherals, we can start considering architectural optimizations to improve system performance. The advantage of using programmable logic to do this is you have more flexibility in what you can do.

A typical microcontroller has a fixed ALU, so performance optimization is limited by the creativity of the software engineer in using that CPU's instruction set. Often this kind of optimization is done to get the last few percent of performance out of a system (by coding in assembly language), and 10%-20% optimizations are hard to come by. With embedded processors running in programmable logic you can still do software level optimizations, but you now have the freedom to take a specific portion of your algorithm and turn it into a block of hardware that is now a part of the ALU – that is, a "custom instruction". Coding software algorithms in hardware can improve performance by hundreds of percent depending on the complexity of the algorithm that is being cast into a custom instruction. The beauty of this approach, however, is that from the software engineer's perspective it's simply another function call. It doesn't matter to him that this function is performed as a hardware block – it's still a function that executes and returns.

Taking this approach further allows us to extend to a full coprocessor such as one with a set of DSP functions. With SOPC Builder this co-processing can be done either off-chip in a separate DSP processor, or as a block of on-FPGA logic. In the case of on-chip logic the exact functions and the way they are optimized can be tailored to each application.

Additionally architectural optimizations such as the number of processors needed for a certain task or the sharing of slave resources with multiple masters and the arbitration of these resources can be experimented with until acceptable performance is achieved. Having a bus-architecture where the slave-resources rather than the bus are arbitrated provides for greater throughput. "Simultaneous Multi-Mastering" and "Slave-side Arbitration" are natural outgrowths of using programmable logic rather than board-level resources for system design.

4.1   Custom Instructions

Custom Instructions provide designers the opportunity to increase system performance by adding new instruction logic to the Arithmetic Logic Unit of their Nios CPU. This technique can be used in any application that has a compute intensive operation that can be translated into a block of hardware. The use of custom instructions gives another axis in design flexibility for the system architect where the decisions to be made are the hardware resources necessary for the custom instruction implementation weighed against the software performance for the same algorithm. Additionally more than one custom instruction can be added to a system as needed.

To get an idea of performance improvements we consider two custom instruction examples. The first, a four-instruction floating point unit, shows the performance increase achieved by replacing the standard floating point C-Libraries with a hardware block that does the same calculation. As a more system-oriented example we look at a software-only MP3 decode routine versus replacing the compute intensive portions with various custom instructions.

4.1.1   Custom Instruction Example: Floating Point Math

Some applications require the use of floating point math. When floating point calculations are done in software, they are notoriously slow. Few microcontrollers provide floating-point support as part of the microcontroller, instead forcing the designer to use either a software library or an external floating point unit. In many cases only a subset of the floating point instructions are necessary for the problem to be solved. Custom instructions allow the designer to choose which operations to implement as hardware and which to leave as software library routines.

To give a concrete example, a subset of floating point operations was created [2] that included the instructions multiply, multiply with negate, absolute value, and negate. This subset floating point unit is 1019 Altera

Logic Elements in size[2]. Each of the 4 instructions is part of a single unit and called with a separate prefix.[3] Table 8-2 shows the performance improvement over the software-only version of each of the 4 floating point instructions.

Table 8-2. Floating Point Unit Custom Instruction vs. Software-Only Performance Comparison

Floating-Point Operation	CPU Clock Cycles		Speed Increase
	Software Library	Custom Instruction	
Multiplication a × b	2874	19	151.3
Multiply and Negate –(a × b)	3147	19	165.6
Absolute |a|	1769	18	98.3
Negate -(a)	284	19	15.0

4.1.2   Custom Instruction Example: MP3 decoder

In many MP3 players, the processor is used for control functions and moving data. A specialized MP3 decoder ASIC is used to do the computation-intensive decode and send data to the sound device [3]. However for this example we started from a full software-implementation of an MP3 decoder and modified it with custom instructions taking benchmark numbers along the way. A software-based MP3 decoder, to achieve minimal adequate performance, needs to run on, for example, at least a 100 MHz Pentium-class processor.

In general an MP3 decoder does the following:
- Reads the MP3 data from storage
- Buffers the MP3 data in RAM
- Decodes the MP3 data
- Synthesizes the MP3 sub-bands into Pulse Code Modulation data
- Sends the PCM data to the Pulse Width Modulator (PWM)

---

[2] This subset floating point unit is approximately the same size, in Logic Elements, of the smallest working 32-bit Nios microprocessor system. This is about 12% of the Logic Elements contained in the Apex20K200E.

[3] Alternatively a separate custom instruction could have been generated for each of the 4 instructions.

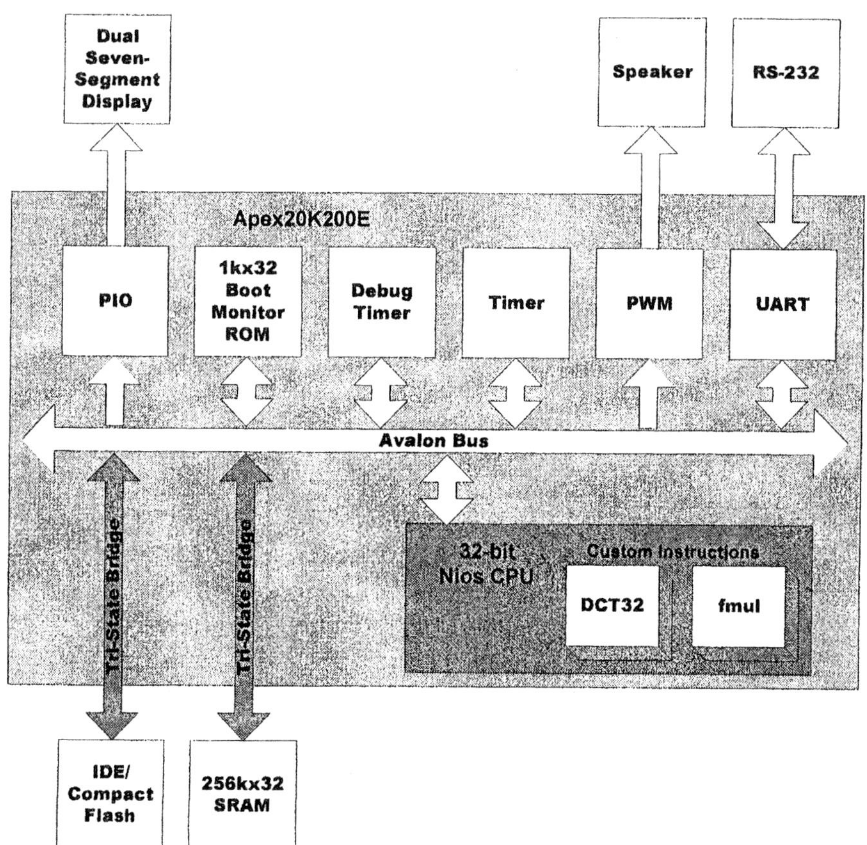

Figure 8-1. MP3 System Diagram

4.1.2.1 The use of Custom Instructions

The majority of the execution time for MP3 decoding takes place in the synthesis of the sub-bands. Therefore, we directed our focus on a function called "mad_synth_frame" (MAD). Here we found two functions that could be easily optimized using the custom instructions, f_mul and DCT32.

<u>f_mul:</u>

f_mul is a macro used by MAD to emulate floating point multiplication using integer multiplication. This macro is defined as:

```
#define f_mul(x,y)(((x)| 0x0001FFFFL )*((y)| 0x0001FFFFL))
```

We note that this function is simply a set of shifts, adds, multiplies and or operations, which are easy to implement in hardware. We created a hardware custom instruction, f_mul, which performs the previous software macro. Below is the code used to define f_mul using our custom instruction hardware:

```
#define f_mul(x,y) nm fmul((x), (y))
```

<u>DCT32:</u>

The function DCT32 is used to perform the discrete cosine transform (DCT) in the MP3 decode. The MAD software uses an optimized DCT hardware block to increase performance. From a software perspective, this optimized algorithm provides a significant performance improvement over the general DCT, using only 80 multiplies where the standard DCT equation requires 1024.

The hardware used for the DCT32 custom instruction was provided by Celoxica [4], a provider of reconfigurable compute solutions with their Handel-C based design tools. Our DCT32 custom instruction was designed with the following features:

- Ability to store 32 inputs and 32 outputs
- During DCT calculation, works independently of CPU
- Uses the prefix instruction to receive commands:
    - Load/unload
    - Start DCT calculation
    - Poll if complete

Because we can poll the custom instruction to see if it is done, we have the ability to run other code in parallel until we need the output of our DCT. At that time, we can poll the custom instruction to see if it has completed the calculation. If complete, we unload the output data while loading the next group of inputs.

4.1.2.2   Performance vs Size for Accelerated MP3

For our test, we measured the number of cycles needed to complete a single mad_synth_frame function using a hardware multiply instruction as well as our custom instructions f_mul and DCT32. The results obtained are presented in Table 8-3. Performance improvement does not come without cost. Custom instructions will improve performance at the cost of additional hardware resources (Logic Elements and memory). The hardware needed for our 3 example designs is also presented in Table 8-3.

Table 8-3. Number of Cycles to Compute an MP3 Frame Using mad_synth_frame

Hardware Used	# Cycles for mad_synth_frame	Additional LEs from Baseline System
HW Multiply only	1,279,000	424
f_mul	293,000	524
f_mul & DCT32 parallel	231,600	1203

The most efficient custom instruction was f_mul, reducing the number of cycles needed by 77% with only an increase in system size of 3%. This small increase in size was due to the optimization of removing the dedicated hardware multiplier in favor of our f_mul custom instruction hardware.

DCT32 run in parallel mode reduced the number of cycles needed by 21% compared to f_mul. This was at the cost of an additional 18.9% of Logic Elements (LE) resources from the f_mul design.

Comparing the extra resources needed for our custom instruction versus the performance increase against our hardware-multiply only baseline system, we see that for a reduction of over 80% in the number of cycles needed to do mad_synth_frame, our system size only increased 22.3%. It should be noted that the percentage increase of overall system performance is heavily dependent on the amount of time spent performing the function that is cast into a custom instruction. If this "inner loop" is the bulk of processing time, then the system performance will improve dramatically, whereas if it is only one of many bottlenecks in the system, then moving it into a hardware custom instruction will have little impact in overall system performance.

4.2 Hardware Co-Processing

The concept of custom-instructions, i.e. using hardware to implement a software algorithm, can be extended more generally into a hardware-coprocessor. In most signal processing systems, designers are faced with the challenge of needing more processing power to compute their complex digital signal processing functions than their processor is able to provide. Using programmable logic, the specific DSP functions needed for an application can be implemented and either utilized by an on-FPGA processor such as Nios, or as a co-processor for a separate dedicated microcontroller. As an example, we implemented an embedded soft-processor and a Finite Impulse Response (FIR) filter co-processor in an FPGA.

The use of signal processing for applications such as 3G wireless and digital video/image processing has grown significantly in the last several years. This growth has resulted in the need for more signal processing ability for improved speed, faster throughput and increased channelization.

In the past ASICs would be developed to address these needs. However, factors like fast time-to-market and high silicon fabrication costs tend to favor other alternatives such as the use of DSP processors with application specific hardware co-processors designed to perform dedicated, computationally intensive signal processing functions.

The use of programmable logic and the extension of a control processor's functionality with a DSP co-processor can speed the development and time-to-market of a project. Being able to profile code, and make architectural tradeoffs as to which parts of the system will reside in hardware and which will be software algorithms allows greater flexibility when considering time and resource allocation in project development.

4.2.1   Code Profiling and Hardware Acceleration using Co-processors

Code profiling involves obtaining the execution time of each function within the user's software code and analyzing each function to get an idea of where the processor is spending most of its time. Core algorithms that consume the majority of the processor's CPU time can be identified after which methods of enhancing the performance of the system are investigated. The example C-code shown in Figure 8-2 illustrates a function that has been designed to run entirely on a single processor.

```
main ()
{ ...variable declarations...
  init();

  while (!error && got_data())

  {
    do_user_interface();
    gather_statistics();
    if (got_new_data())
      Tform (in_buf, out_buf);
    check_for_errors();
  }

  cleanup();

}
```

Figure 8-2. Example C-code and code profile of an arbitrary DSP function called Tform

Profiling the code points to a single bottleneck where we find a function called Tform[4] taking up most of the CPU's time. This software bottleneck degrades overall performance but can easily be overcome with the use of a custom hardware accelerator which will perform the Tform function. As we saw in previous examples, we can offload the "heavy-lifting" required by the

---

[4] The function Tform is used in this context to represent some arbitrary DSP function like a transform or a filter.

Tform function into a custom hardware co-processor allowing the CPU to spend most of its time on general purpose control functions.

The performance improvement afforded by offloading this computationally intensive signal processing task to a custom hardware co-processor can be significant as illustrated in Figure 8-3, especially if the function that is causing the bottleneck is utilized repeatedly within the target application.

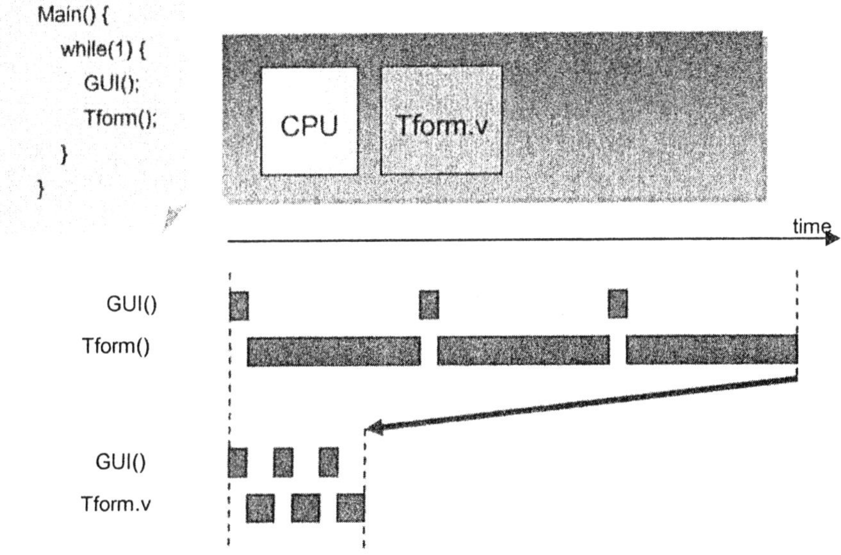

Figure 8-3. Performance improvement achieved by implementing custom hardware co-processors for "heavy-lifting"

4.2.2 Parallelism in a Sequential Environment

One of the key advantages of using hardware accelerators is the employment of parallelism, where multiple tasks are executed simultaneously. In Figure 8-4, we see how the Tform task execution could change with the addition of two more Tform co-processors in the system and provide additional performance improvements. The entire computation of the Tform task could now be divided between three Tform co-processor hardware accelerators. Each Tform co-processor only runs for a fraction of the total execution time required to complete a single Tform computation task. However, these tasks are running in a sequential nature and do not take advantage of parallel execution. Hardware functions that are not data-dependent could be executed in parallel, resulting in an overall drop in utilization and improved performance.

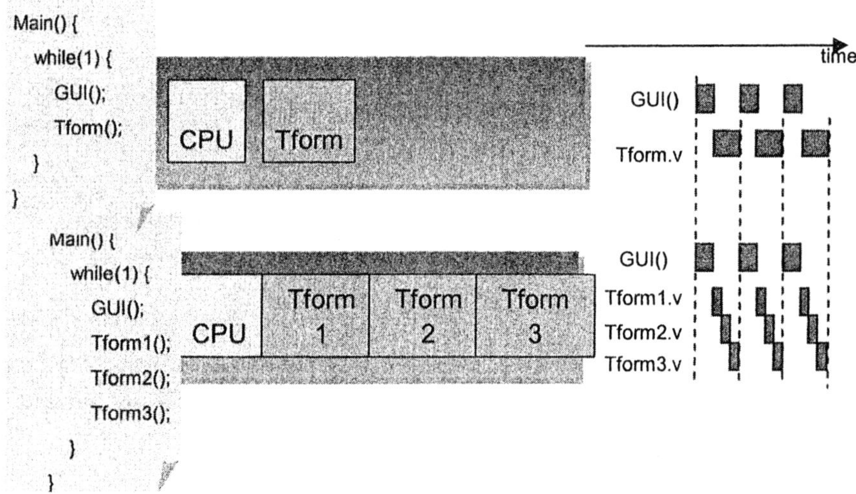

Figure 8-4. Implementing multiple Tform co-processor hardware blocks

But how do we provide parallel acceleration in a sequential environment such as a program running on a microprocessor? One solution that enables software to perform multiple tasks at once involves the use of threads. The term thread refers to a section of code executed independently of other threads within a single program [5]. While threads executed on a single CPU are still executed sequentially and thus do not affcct thc actual parallelism of the execution, threads used in conjunction with co-processors provide the ability for the processor and all co-processors to operate in parallel.

Modifying the example such that it calls the various Tform functions in threads enables the parallelism illustrated in Figure 8-5. The amount of parallelism and utilization afforded by the use of threads is governed by the critical path of the dataflow through the various functions implemented in the design.

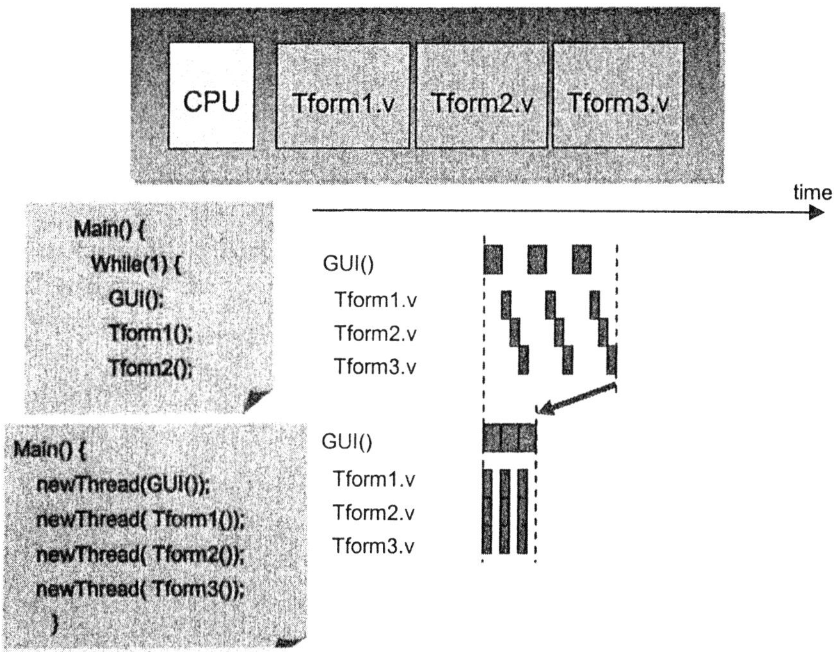

Figure 8-5. Parallelism provided by the use of threads

4.2.3 The Hardware Interface

The co-processing function Tform could be a piece of custom logic or pre-configured off the shelf IP like a FIR filter, Viterbi or Turbo encoder/decoder. A generalized view of the Tform function is shown in Figure 8-6. The memory interface of the hardware coprocessor is just as important as the functional implementation of the Tform accelerator. To optimize data throughput, we sandwich our Tform function between two DMA peripherals. A DMA peripheral allows for data transfers between two memories, between a memory and a peripheral, or between two peripherals and is used in conjunction with streaming-capable peripherals, allowing data transfers to occur without intervention from the CPU [6].

In Figure 8-6 DMA1 is used for data accesses that feed the input of the co-processor. The read master of DMA1 is fed by a memory source (on-chip, off-chip, or from an Analog/Digital Converter (ADC)). The output of DMA1 is connected directly to the input of the Tform co-processor. DMA2 is used to access the output data of the co-processor and feed it to a memory buffer to await further processing.

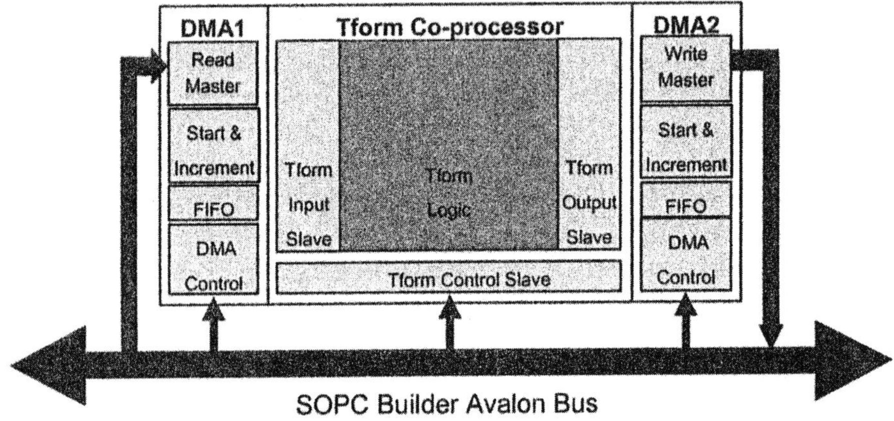

Figure 8-6. Co-processor interfaced to the Avalon bus via two half-DMA peripherals

The read and write half-DMA peripherals are components available in SOPC Builder. The Avalon bus is the non-proprietary bus specification used by SOPC Builder to wire up peripherals and the accompanying logic it generates. It is an interface that specifies the port connections between master and slave components, and specifies the timing by which these components communicate. It also supports simultaneous multi-mastering where multiple masters are allowed to perform bus transactions at the same time, as long as they do not access the same slave during the same bus cycle [7]. In the event that multiple masters attempt to access the same slave at the same time, a built-in arbitrator prioritizes accesses to that slave.

4.3   Architectural Optimizations

To demonstrate this inherent flexibility for performance optimization incorporating custom hardware and architectural optimizations an embedded web server application was chosen because it is a relatively common embedded task encompassing several complex software components such as file system support, TCP/IP stack, and Ethernet interface. The speed at which the server is capable of servicing requests for data was chosen as the performance benchmark for each design optimization.

4.3.1   Baseline System

The baseline from which we start will be the 32-bit Nios RISC processor core running at a system clock speed of 33MHz. Although fmax of the baseline system was significantly higher, this clock speed coincides with clock speeds of many traditional 16-bit and 32-bit microcontrollers. In addition to the CPU, a simple set of peripherals completed the hardware portion of the baseline system, including:

- ISA Interface to a 10base-T Ethernet MAC
- Shared program & data memory
- Host communication via UART
- A timer peripheral to measure the web server's performance

Software running on the system consisted of a basic web server capable of responding to simple HTTP requests, as well as simple CGI requests for things such as dynamic HTML. In addition, the web server was designed to read a timer peripheral before, during, and after servicing an HTTP request to log throughput calculations, which are then displayed on a dynamically generated web page. A very simple read-only file system was implemented using flash memory to store static web pages and JPEG images that would later be served. This top-level software was identical throughout the performance enhancement project, as it was essential to provide identical application-layer software support.

Once the server was up and running the host PC accessed several JPEG images of varying sizes stored in the file system. For each of these requests two throughput calculations were performed:

1. Transmission throughput, which reflects the latency between starting to send the first TCP packet containing the HTTP response until the file was completely sent. The transmission throughput could theoretically reach a maximum of 10Mbps since that is the speed of our external MAC. This may also be thought of as the raw network speed that the CPU and TCP/IP stack are capable of sustaining. It should be noted that a substantial portion of each packet – the bytes making up the Ethernet and TCP headers - were ignored for this calculation, and only the payload taken into account.

2. HTTP server throughput, which takes into account all delay between the incoming HTTP connection request and file send completion. This calculation includes the transmission latency above, but also measures the time the HTTP server took to open a TCP connection to the host, and subsequently search for and obtain the requested file from the file system.

With these desired measurements in mind, the web server application was put to the test to serve up JPEG images of varying sizes. During each transfer several snapshots of the timer peripheral were taken. Table 8-4 illustrates the performance achieved with the baseline system serving several JPEG images across the local area network[5] to a host PC.

Table 8-4. Baseline system performance

File size (bytes)	Transmission latency (ms)	Transmission throughput (Mbps)	HTTP server latency (ms)	HTTP server throughput (Mbps)
14650	38	3.08	57	2.06
37448	96	3.12	142	2.10
60036	153	3.14	226	2.12

4.3.2   Areas for Optimization

To assist in visualizing the data flow for each optimization, a basic block diagram was drawn. Figure 8-7 shows dataflow between peripherals in the baseline system. In this example, the Nios CPU's data master port is used to read data memory (SRAM) and write to the Ethernet MAC. This would occur for each packet transmitted in the baseline system.

---

[5] Performance across a cross-over cable directly to the host PC was nearly identical.

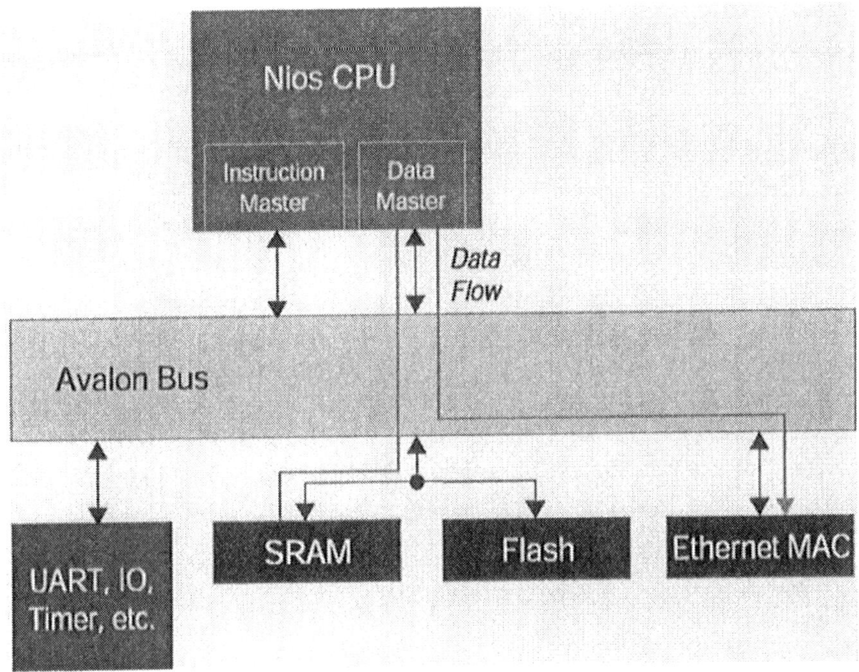

Figure 8-7. Baseline system dataflow during packet transmission

With the above process in mind, we can explore several approaches for performance optimization of the design. Specifically we will be looking at using a DMA to transfer data from incoming packets into memory without the intervention of the microprocessor, the use of a custom peripheral to do the checksum calculation, and the combination of these two. Additionally we will be looking at optimization of the slave-arbitration priority for the memories to provide maximum data throughput.

4.3.3   Dataflow Enhancement with DMA

Adding a DMA controller to our example system offers the ability to free the CPU of mundane data transfer tasks between peripherals. In addition to reducing the latency required to shuttle data about in the system, adding DMA to an embedded system in programmable logic affords another key benefit: the ability to fine-tune bus architecture. Traditionally with discreet components, printed circuit board (PCB) layout is determined up front with little room for any sort of bus architecture change. This leaves only fine-tuning of software and faster clock speeds to drive up performance of a system near the end of a design cycle. Designing with a soft-core microprocessor and programmable logic allows greater leeway in this respect – while aspects such as the data and program memory PCB traces are fixed, bus arbitration schemes for sharing memory between DMA and CPU are configurable. Again, this architectural flexibility comes because we implement bus interconnects inside the FPGA, whose routing resources can accommodate data flow between logic elements and memory arrays throughout the device.

To showcase the performance difference of simply adding a DMA controller, we kept the existing shared program & data memory, and allot the CPU slightly higher priority for any conflicts with the DMA controller[6]. During DMA operation, the Nios CPU is free to access other peripherals as the bus interconnections are physically separate. For access to the shared SRAM during DMA operation, arbitration is performed. Figure 8-8 shows dataflow between peripherals during packet transmission after DMA was added.

---

[6] The Nios CPU was allotted four times the access than the DMA controller to shared data/program memory.

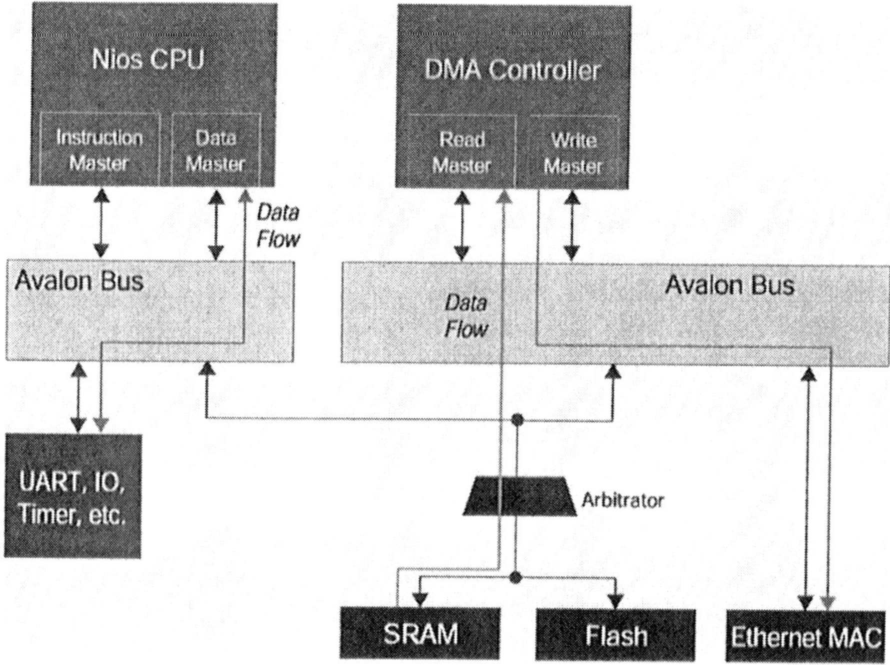

Figure 8-8. DMA system dataflow during packet transmission[7]

After adding the DMA controller to the system, software modifications were performed in a few key areas where the CPU previously transferred data. Using DMA to transfer packets between Ethernet MAC and data memory is perhaps the most obvious place to start. Without DMA the CPU must sit in a while-loop, to transfer all data for a sent or received packet between the outside world and data memory. This task is easily shifted to a DMA transfer instead. Analysis of the TCP/IP stack source code, by following the packet assembly process, revealed two additional loops where the CPU copies payload data for a packet under assembly. These routines were also handled using calls to the DMA controller.

After adding the DMA controller and modifying our software library we're ready to test our server's performance once again. The benchmark was performed in an identical manner to the baseline system with the system clock still set to 33Mhz. Table 8-5 summarizes the results.

---

[7] The DMA controller's control slave, which the Nios CPU uses to access DMA control register, has been omitted for clarity.

Table 8-5. Baseline system with DMA Performance

File size (bytes)	Transmission latency (ms)	Transmission throughput (Mbps)	HTTP server latency (ms)	HTTP server throughput (Mbps)
14650	18	6.50	24	4.88
37448	44	6.81	58	5.16
60036	71	6.76	94	5.10

Notice that we've seen a substantial improvement – transmission throughput is, on average, just over double that of the baseline system. The entire HTTP server throughput is about 2.5 times that of the baseline system. Suddenly we've begun to really use our little 10base-T Ethernet MAC!

Since the DMA controller itself was included as a license/royalty-free component of SOPC Builder, the hardware design cost was zero, and took less than an hour of engineering time. But nothing comes without its cost. The DMA controller and its bus connections to Ethernet MAC and data memory use additional FPGA resources. After DMA was added to the system, the total logic usage for our system climbed to just over 3,600 LEs – a 36% increase in logic usage. Also, the software design changes took a one day effort for one engineer to add the DMA support to the relevant TCP/IP library routines. For any project the cost/time tradeoffs must be analyzed. From the baseline system, with the application of 1 day of engineering time and roughly 900 LEs, the system performance improved by 250%.

4.3.4 Shifting the computation burden: A custom peripheral

Adding a DMA controller to the system, while straightforward, is by no means the only method with which we can enhance our web server's performance. Once again, looking at the process by which we serve up data from our web server reveals another compute-intensive loop which the CPU must execute for every packet it transmits: The TCP checksum.

Checksum calculations can be regarded as a necessary evil in dataflow-sensitive applications because of the latency they induce. A brief analysis of the TCP checksum calculation on a maximum-sized (approximately 1,300-byte payload) packet showed that it takes approximately 33,000 clock cycles to step through the data adding it together to produce a checksum. At a 33 Mhz clock speed it equates to approximately 1ms latency for each maximum size packet that we send out. In our benchmark, the largest file (60KB) breaks down into just over 46 maximum-sized packets. That's 46ms our of the 156ms transmission latency in our baseline design!

The inner loop of the TCP/IP stack checksum, which performs a 16-bit one's complement checksum calculation, is shown in Figure 8-9. Other portions of the function that add up information for packet header fields and do some additional byte swapping have been omitted – the following simply runs through the entire packet payload and sums it all up:

```
while(byte_count > 0)
{
    x = nr_n2h16(*w++);    //Byte swap routine

    if(byte_count == 1)    //If we're at the end of packet
        x &= 0xff00;       //mask off checksum of finalbyte

    result += x;           //Add data to partial checksum

    byte_count -= 2;       //Continue 16 bits at a time
}
```

Figure 8-9. Example "Inner Loop" of software-based checksum

Adding up data repeatedly is a rather simple task for hardware, so we can reduce the CPU's instruction burden by performing this routine calculation in hardware. To accomplish this task a Verilog implementation of the above routine was designed. The checksum peripheral operates by reading the payload contents directly out of data memory, performing the checksum calculation, and storing the result in a CPU-addressable register.

When all was said and done, a few pages of Verilog code had been created that performed checksum addition of 32 bits of data on each clock cycle. This worked out to approximately 386 clock cycles to compute the checksum of a 1,300-byte packet payload, a speedup of nearly 90 over the software algorithm!

Figure 8-10 shows the dataflow between peripherals with the addition of the custom checksum peripheral during checksum calculation. Like the DMA example, arbitration is only performed for access to the shared peripheral (SRAM).

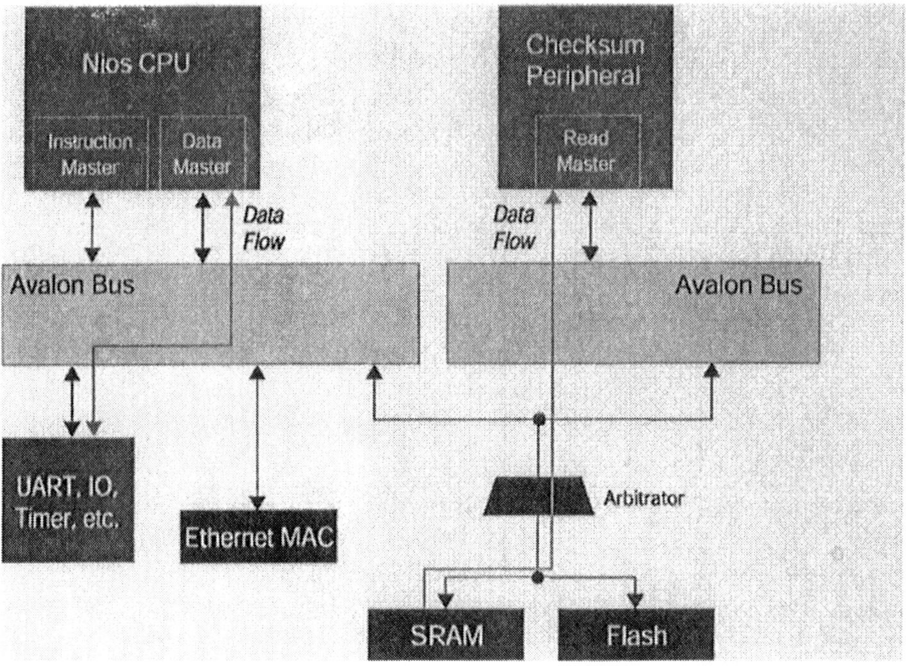

Figure 8-10. Dataflow diagram between peripherals including custom checksum peripheral.[8]

Once the system was re-generated and the benchmark loaded, the performance boost (Table 8-6) was in line with our original expectations. Recall that for the 60KB image, our calculations showed that roughly 46ms of latency is required for checksum calculation. In the benchmark on the same 60KB image, transmission latency decreased by 44ms, and overall latency by 45ms which can be attributed to the checksum engine's latency and software setup for each calculation.

Table 8-6. Baseline system with custom checksum peripheral performance

File size (bytes)	Transmission latency (ms)	Transmission throughput (Mbps)	HTTP server latency (ms)	HTTP server throughput (Mbps)
14650	27	4.37	45	2.60
37448	68	4.40	114	2.62
60036	109	4.40	181	2.65

[8] The checksum peripheral's control slave, which the Nios CPU uses to access peripheral registers, is omitted for clarity.

The net benefit of this addition works out to an average transmission throughput increase of 40% and average HTTP throughput increase of 25% over the baseline system. Like the addition of DMA, the custom peripheral had its own cost in development time and logic utilization. Our system with the checksum peripheral now consumes about 3,250 logic elements – a 22% increase over the baseline system. Since the hardware portion of the design provided a clean interface to CPU via control & status registers, software development & debug time was on the order of a couple of hours! The real engineering effort went into the creation of the peripheral's Verilog source code, which took about 3-4 days of development and debug time.

4.3.5   Putting it all together

Using the above optimizations we can now add them to get a combined improvement to the performance of our example system. The final design run uses the baseline system, DMA controller, and custom checksum peripheral together. All hardware and software changes to the system were identical to the above steps. Table 8-7 summarizes the results of combining all system optimizations.

Table 8-7. Performance with both DMA and checksum peripherals added

File size (bytes)	Transmission latency (ms)	Transmission throughput (Mbps)	HTTP server latency (ms)	HTTP server throughput (Mbps)
14650	12	9.76	20	5.86
37448	31	9.66	49	6.11
60036	51	9.42	79	6.07

Table 8-8. Webserver Throughput Performance Comparison

Architecture	HTTP Server Throughput (Mbps)	Transmission Throughput (Mbps)	LE Usage
Baseline	2.12	3.14	2700
With Checksum	2.65	4.40	3250
With DMA	5.10	6.76	3600
With Checksum and DMA	6.07	9.42	4400

Table 8-8 compares each of the optimizations for the best cases (largest file size sent over Ethernet). The ten megabit limit has indeed been reached. In comparison to the baseline system, transmission throughput has increased over 300%, while HTTP throughput has increased by over 280%! The total logic utilization for the system with both DMA and custom checksum logic has risen to just under 4,400 LEs, a 65% increase from the baseline system.

5. THE FUTURE OF SYSTEMS ON A PROGRAMMABLE CHIP

The era of programmable logic has just begun and the future looks bright. All the tools and component pieces are available to quickly and reliably create useful systems on a programmable chip. That's not to say there won't be more improvements along the way – on the contrary. We're in a period of time of rapid evolution for the SOPC tools, IP, and software. If the world of software development is a guide it will take several years of continued infrastructure development to mature the product offerings for SOPC development. In that time, however, other areas will be changing in parallel.

As the process technology continues to smaller and smaller geometries, the raw performance of SOPC systems will improve – without an expensive step of redesigning your system for the new process technology. Said another way, when a design moves from 0.13 μm to 90nm, an SOPC design simply must be recompiled to target this new FPGA to take advantage of the improved performance. The time and costs associated with developing, testing, and manufacturing an ASIC do not stand in the way of bringing a SOPC product to market.

Other technological changes we anticipate in the world of FPGAs include additional dedicated hardware blocks, new fabrics on new process technologies, interfaces to external processors, and the merging of hardware and software algorithmic development. Each of these developments will evolve over the coming years and will provide some level of performance gains for system designers either in raw performance or time-to-market or both. As stated before, FPGAs naturally benefit from process technology improvements; changing the overall FPGA architecture may also offer improvements for certain classes of problems in the same way that the change from CISC to RISC architectures provided improvements to software developers in the past. But in some cases one needs as much raw general-purpose microprocessor processing power as is available at the time. In these cases, having an interface to these processors so the FPGA can be used to offload tasks from the main processor or augment its performance is a natural extension of the SOPC concept. And eventually the question of software vs. hardware will be made moot as languages and tools develop that allow the fluid flow from a software algorithm to a hardware block without any intervention from the system designer.

The times are exciting for system designers. Systems on a Programmable Chip using FPGAs offer them tools they have never had before. The continued evolution of these tools will provide for faster experimentation and implementation of ideas than was ever possible before.

REFERENCES

1. SOPC Builder Design Flow & Features, found at the web link
    http://www.altera.com/products/software/system/products/sopc/design/sop-design_flow.html/
2. AN 188: Custom Instructions for the Nios Embedded Processor, Altera Corporation.
3. http://www.mp3projects.com/beginners.html - two/
4. Celoxica, 20 Park Gate, Milton Park Abingdon Oxfordshire, OX14 4SH LTD United Kingdom. URL: www.celoxica.com.
5. S. Oaks and H. Wong, *Java Threads*, 2nd edition, O'Reilly, 1999.
6. *Nios Embedded Processor Peripherals Reference Manual*, Altera Corp., January 2002.
7. *Avalon Bus Specification Reference Manual*, Altera Corp., July 2002.

Chapter 9

STAR-IP CENTRIC PLATFORMS FOR SOC
ARM® PrimeXsys™ Platform Architecture and Methodologies

Jay Alphey, Chris Baxter, Jon Connell, John Goodenough, Antony Harris, Christopher Lennard, Bruce Mathewson, Andrew Nightingale, Ian Thornton, Kath Topping
ARM Ltd

Abstract: We describe the use of star-IP core-based subsystems as the cornerstone of a platform-based design paradigm. An ARM platform is an instantiation of a set of carefully market-targeted architectural-decisions encapsulated in an embedded and configurable subsystem consisting of an ARM core, AMBA™ communications fabric and a ported operating system (OS). Around this pre-specified sub-system, a derivative-product development-package is supplied. This development package provides for configuration and extension of the platform during the creation of an optimized and differentiated system-on-chip (SoC) design. We describe the structure of this development-package, and its foundation in a set of mutually consistent model-views of the platform design. Each platform model provides the speed and visibility required for specific SoC development tasks: hardware integration and development, hardware dependent software development, application software development, and system verification and validation. In this chapter we describe both the theory of platform support, and a specific ARM instantiation of this: the ARM1136JF-S™ PrimeXsys Platform.

Key words: ARM, AMBA, PrimeXsys Platforms, Star-IP, Modeling, Verification and Validation, Standards, Interfaces

1. CORE-BASED ARCHITECTURES

This chapter describes the conceptual framework supporting ARM's platform-based design solutions, and relates this to a deliverable ARM product: the ARM1136 PrimeXsys Platform [1]. The authors of this chapter are the lead technical and product architects of ARM's platform solutions. ARM Limited is a well-established provider of 16/32-bit embedded RISC microprocessor solutions for embedded-systems design, of which platforms are a cornerstone. ARM was founded in 1990, and rapidly established industry leadership in licensing of high-performance, low-cost, power-efficient RISC processors to international electronics companies. ARM extended this intellectual-property (IP) portfolio to better enable customers' SoC design experience. Initially, this extension focused on providing a broad set of peripherals through the ARM PrimeCell® library, and more recently this has matured into supply of full SoC platform solutions around the ARM PrimeXsys Platform product family. ARM's value proposition is not built solely on the excellence of its design IP, but also in provision of IP integration support. For our platforms as well as our core products, we supply comprehensive models, debugging and hard-prototyping environments, extendible verification and validation IP. To further enhance integration ease of ARM IP, ARM is an active contributor to, and provider of, open standards. ARM is extensively involved in the Virtual Socket Interface Alliance (VSIA^{TM}) [2], Open SystemC^{TM} Initiative (OSCI) [3,4], Accellera [5] and is the provider of the AMBA Bus [6] interface standards.

In this chapter, we explain the layered approach to SoC platform based design adopted by ARM, both for internal usage and for support of external customers. The word 'layered' implies that several distinct, but functionally consistent, views of the platform are built and maintained during the SoC architectural-design and implementation process. ARM supports three specific platform layers, as depicted in Figure 9-1. At the lowest layer, the hardware layer (Platform Layer 1), there is a micro-architectural representation of the compute-engine - the interconnected hardware that runs the embedded software; e.g., the register-transfer level (RTL) description of an SoC. Above this is the SoC-integration and middleware layer (Platform Layer 2) that configures the operating-system (OS) to the hardware architecture and encapsulates the hardware-dependent software layer; e.g., specific OS port with memory and peripheral drivers. Finally, there is the software that targets the platform to a specific domain, the application software layer (Platform Layer 3); e.g., a software stack performing graphics acceleration for the multimedia space. The adopter of the platform has the ability to configure the platform at any of the implementation layers. There are several ways in which this custom configuration may be achieved: by providing application-oriented software IP; by extending the platform with application specific design IP that may take the form of: memory-subsystems, custom peripherals, digital signal processing (DSP) accelerators; and by configuring the platform itself in terms of arbitration policies, bus hierarchy, and memory maps.

Integration of the custom-configured platform into a product may be viewed as a final or fourth layer that is built on the three-layer platform support. Platform based design provides the IP consumer with rapid access to the product-layer, improving time-to-market (TTM) whilst providing suitable customization options for product differentiation.

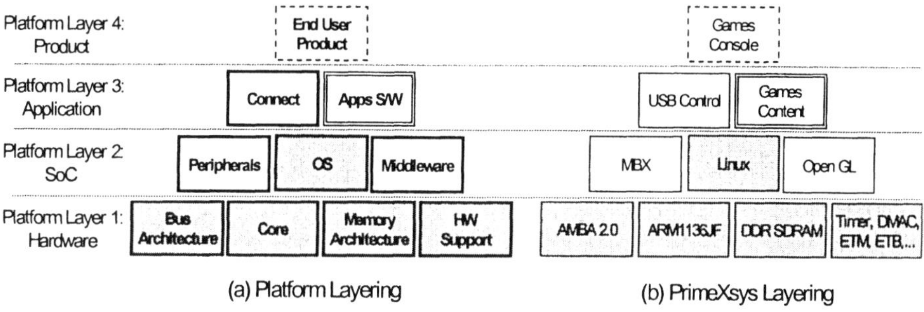

Figure 9-1. Conceptual Platform Layering

In Figure 9-1, the abstract platform layering concept is depicted on the left-hand side, with an example 'customer' configuration of the PrimeXsys Platform on the right. Against each of the platform layers introduced above, the main design elements of that layer are defined. At Platform Layer 1, hardware-integration, the critical elements in the definition of a platform are: the bus architecture, including bus-hierarchy, bridging and interface protocol; the core support, both for control and signal processing; the memory architecture, including the local and global connectivity matrix, memory controller policies, caching policies and memory-space mapping; and the hardware support for the SoC, including such components as trace and debug structures (cross triggering, trace-buffering, etc.), watchdog timers, and direct-memory access (DMA) support. At Platform Layer 2, the SoC integration layer, the IP involved is the OS ported to the hardware layer, the hardware-dependent software drivers and the peripheral bundles (hardware object for integration to the hardware layer, provided with a driver that may be configured to the OS and system memory map). At Platform Layer 3, the software development layer, are the complex stacks involved in control of SoC communication (e.g., a telecommunications communications stack), and the application software that programs the SoC to product intent (e.g., PDA user tools like calendars, etc.).

The most essential elements of a platform are depicted in Figure 9-1 as the gray boxes, with the platform extensibility and configurability shown by the white solid boxes. ARM's platform products are defined with a particular configuration of the IP in the gray boxes, and the design is rapidly extendable though use of ARM's PrimeCell® or 3rd-party AMBA-compatible peripheral libraries.

As a platform provider, it is not sufficient to purely offer a platform implementation and a compatible IP library. The benefits of platform-based design only emerge through comprehensive exposure and support of the platform's configurability and extensibility. Customers require a guarantee of efficient IP integration, and that an efficient representation or 'view' of the design supports each platform layer. Each view of the platform must provide sufficient design resolution for the development requirements of a platform layer, and sufficient execution-speed (i.e., equivalent cycles-per-second) to comprehensively validate architectural and design decisions for that layer. Primarily, the representations that must be provided in support of a platform are views for: hardware integration, SoC hardware/software validation, SoC architectural exploration, and application software development (hard and soft prototyping).

ARM provides the following design infrastructure in support of platform-design views:

1. interfaces: standardized interfaces for design, verification and model IP that support product integration at each of the three platform-design layers.
2. models: core and platform.
3. software development support: hard and soft prototyping with a debugging environment that supports re-targeting.
4. verification/validation (V&V): reusable components and methodology, and interface protocol compliance checkers.

This chapter introduces the reader to the concepts of core-based platform support based around the four key aspects listed above. We begin with a discussion of the principles of platform IP exchange (Section 2), and the need for standardization to enable IP and platform reuse (Section 3). We then proceed with a description of the support required for the platform layers: Platform Layer 1: hardware development and integration (Section 4), Platform Layer 2: hardware-dependent software development (Section 5), and Platform Layer 3: application software development (Section 6). In parallel with the design process, verification and validation is supported (Section 7). As ARM does provide all these facilities in support of our platform products, we describe a specific platform product as an example: the ARM1136 PrimeXsys Platform (Section 8). Finally, the chapter closes with a set of conclusions (Section 9).

2. ADOPTING A PLATFORM: THE USER VIEWS

2.1 Introduction

Adoption of a platform requires the communication of design requirements and implementations between multiple users. Models of a platform are critical for enabling this communication. As an example, consider embedded software development kits: the embedded OS company Symbian® supplies a tool-kit for developing applications for Symbian OS [7], but the execution environment on which you develop with this kit is your desktop PC. There is no guarantee that software developed within this environment will directly port to the final product unless there has been a model of the platform on which to execute the software. A platform model provides an executable description of the SoC architecture enabling a high-degree of debugging (better than physical designs), and the flexibility to explore the interplay between software and SoC design requirements. In this way, modeling enables technical communication between the embedded software development teams, and the SoC architectural specification and design teams.

In this section, we will describe the users and adopters of the platform through the design flow from system-design to physical implementation, exposing the required communication and IP exchange channels for assembling of a complete embedded system. Throughout the remaining sections of this chapter, we then build upon these requirements to provide a methodology that supports the technical and business constraints of complex IP exchange.

2.2 Exchange of Platform IP

Platform IP deployment is often seen as a one way street where the IP creator licenses (hopefully) high quality IP to a licensee who integrates this into their design. Simple IP is delivered with perhaps implementation code, simulation models for RTL design and integration guides documenting the function of the platform. For simple hardware-centric components whose functionality is well understood and for which there is little or no software included in the package, this often suffices.

In the more complex case of platform IP, this simple view of IP licensing does not sufficiently capture either the complexity of the relationships between the creator and the user(s) of platform IP, or the difficulties in transferring different views of platform IP between creator and the user(s) of the IP. Required communication between the parties is often complicated by the need for concurrent development of a product.

2.2.1   Platform deployment roles

The deployment of platform IP for multiple licensees who, themselves, target yet more end users is best achieved through use of a streamed, and possibly concurrent, approach to the development of a product. We identify three distinct roles: the IP Creator, the IP Licensee and the IP User.

- IP Creator: creates the base platform and ports suitable operating systems to facilitate the exploitation of the functionality of both hardware and software by the licensee and end user;
- IP Licensee: extends the platform through differentiation, targeting the base functionality to specific application domains, adding their own hardware and software IP;
- IP User: integrates the differentiated platform into a product, making use of the wide range of third-party support attracted by the base platform.

The organizational relationship of each of these three contributors to the deployment of IP has significant bearing on the requirements on that IP. In Figure 9-2 we identify a chain of IP deployment in which each of the three roles can be represented by separate companies or different parts of the same organization:

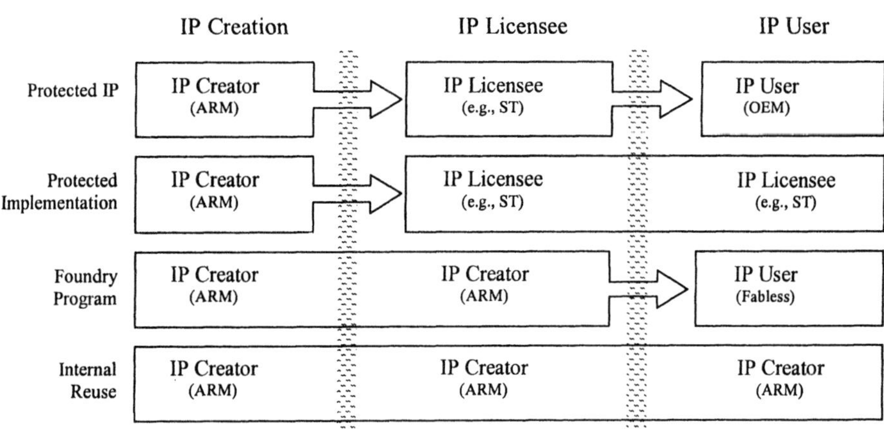

Figure 9-2. Three roles in IP deployment

Wherever exchange of IP involves separate organizations, the security of the IP itself becomes important as the IP Creator is likely to have no control over the IP Licensee's choice of IP User. The IP Creator must therefore provide a number of discrete but consistent views of the IP that enable the design processes to be employed by a wide range of IP Users. For example in the case of a simple hardware component, these 'black-box' views may consist of: a compiled cycle-accurate model for RT design; a compiled phase-accurate model for netlist integration; and a power-model, timing-model, floor-plan and routing-constraint model for physical integration. The set of views which must be provided for complex subsystem IP that includes hardware, software and verification support is significantly more extensive than that required for basic hardware IP, and this creates a significant increase in the scale of the problem for successfully deploying a platform.

2.2.2 Design process interaction

The end users involved in platform IP deployment do not work with that IP in isolation. For a typical product, the IP Creator will create a base platform; the IP Licensee extends that platform adding unique value that attracts customers; and finally the IP User will deploy the differentiated platform in a product.

For IP products to be successful, they need to be attractive to as wide a range of licensees and end users as possible, but in the initial stages of specification and design, an IP Creator and IP Licensee will typically work together with strategically important IP Users to prove the value of that platform. The user interactions depicted in Figure 9-3 identify a number of exchanges of design representation throughout the development of IP. These interactions correspond to well-defined stages in the SoC implementation process, as indicated by the arrows between the idealized top-down development flows [8].

In development of their part of the completed product, each user in the design chain will seek to define their system architecture and then decide how that architecture is best represented by hardware and software components. The correct interaction of those hardware and software components must be proved during the design stage before implementation decisions are made. For the ARM PrimeXsys Platforms, each base platform is provided together with operating system ports and the 'correct interaction of hardware and software components' will include the porting of device drivers and booting of ported operating systems. We describe how this is achieved in Section 5 of this chapter.

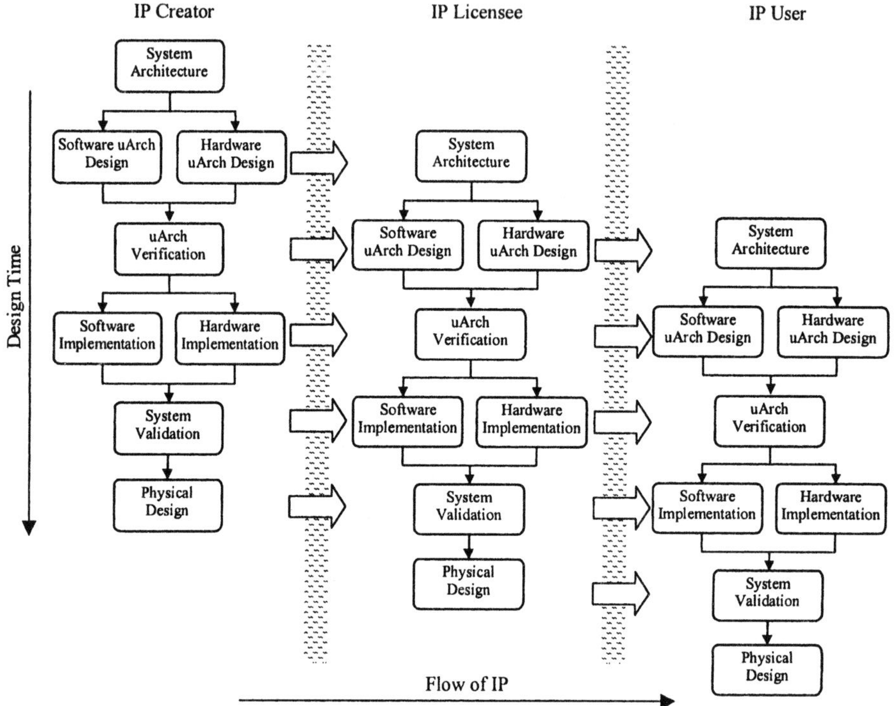

Figure 9-3. Platform IP user interactions

Consider the context of interaction between each of the users in the deployment of the platform, Figure 9-3. The results of the IP Creator's micro-architectural investigation will be used as the basis of the architectural decisions of the IP Licensee. In turn, the specification of the licensee's differentiated platform will then be used by end users (IP Users) to define the micro-architecture of their products.

Each view of the platform reflects the same system architecture, and designers can use test software in any of the higher-level views, providing a high degree of confidence in the design prior to tape out. This provides a valuable environment in which to investigate system bandwidth and performance requirements. System views must be extendible, allowing designers to exploit the advantages of a well-supported, pre-verified base platform of hardware and software IP, whilst differentiating their own application with their own IP.

Additionally, there will also be a transfer of validation IP. This is critical to the success of the platform since the recipient of the platform IP will need to ensure that any extensions they add to the base platform and to the differentiated platform do not invalidate that core functionality. Were the licensee unable to prove that his differentiated platform still implemented the base functionality of the licensed platform IP, the IP User would be unable to exploit the wide variety of third party tools that are attracted to a standard base platform. How a consistent and reusable verification and validation methodology is achieved for complex hardware and software IP is described in Section 7 of this chapter.

3. PLATFORM INTERFACES: STANDARDS

3.1 Introduction

Platform, and design and verification reuse in general, hinge upon the existence and broad adoption of interfacing standards. The obvious standard critical to IP integration is a bus interface, for which ARM has created the AMBA open-standard framework [6]. In this section, we describe the mechanisms by which ARM supplies and contributes to industry standardization to better enable IP integration.

Standard bus interfaces are critical to efficient hardware IP integration, but these are only one of several well-qualified interfaces that a platform must expose. We first describe the general open-interface issue as it relates to platforms, then address more specifically hardware IP communications channel (e.g., AMBA bus) that ARM platforms support.

3.2 ARM standards generation and engagement

There are three different types of standards that ARM provides or engages with. Firstly, there are the product-deliverable or 'platform standards' that may include ARM-proprietary interfaces. Secondly, there are the two forms of open-standards with which we engage: ARM-community (e.g., AMBA) and open-committee (e.g., Accellera, OSCI, IEEE, VSIA) standards. We describe each of these in more detail here.

Platform standards describe the requirements for interfacing IP (hardware and software design, models, and debugging environments) to our platform design, and its soft-prototype and hard-prototype instantiations. Platform standards apply across all ARM platform products. The ARM platform customer may choose to structure their internal design and verification flows to conform to these ARM platform standards, thereby assuring compatibility with ARM's future platform products. There are eight basic platform product standards, described in the following subsection. Some of these interfaces are private and provided with the platform deliverables (e.g., debug cross-triggering requirements), some are ARM-community standards (e.g., AMBA 2.0), and some adopt the principles of the open-committee standards (e.g., Virtual Component Transfer (VCT) Specification of VSIA).

Open standards provide common structures across the electronics design industry that enable IP and tool providers to support the IP integrator. The output of open standardization efforts must be attainable and usable in a non-restrictive way. The difference between open-committee and ARM-community standards is the mechanism of standards creation. Both styles of open-standards structures must exist, and both play an important role in enabling platform products.

The ARM-community standards are of foremost importance to ARM. These express the formats, interfaces and packaging choices to which all our IP products are offered. ARM-community open standards ease basic integration issues by ensuring an IP and EDA-support base that offers a broad range of solutions to ARM customers. These standards describe our lead products, so are driven by firm time-to-market requirements. The ARM-community standards are jointly developed by a representative group of ARM partners. These are comprehensively validated through reference implementation before release, then opened for free to the public under a simple non-restrictive license. These licenses are materially similar to the AMBA 2.0 license agreement. This approach provides a good balance between support-breadth, speed of generation, and clarity of ownership.

Besides driving ARM-community standards, ARM commits significant time and effort into engagement with open-committee standards. The emphasis of open-committee standards is broad-based consensus rather than product-support time-to-market requirements. Broad-base consensus is important for industry-wide standardization of a maturing technology or the encouraging of standardized interfaces and formats that affect ARM-IP adoption indirectly. These open-committee standards also play an important role in closing the gap between a set of industry de-facto standards that have become non-differentiating. Through active encouragement of open-committee standards, ARM is ensuring that platform delivery is possible into multi design-language environments (e.g., Verilog / VHDL co-simulation support), and multi-language validation environments (e.g., standard interfaces for verification components, and consolidation of in-line assertion languages).

3.3 Platform integration standards

In the following table, Table 9-1, we describe the eight platform standards to which platforms should be provided. This list covers a comprehensive set of concerns for customer integration of IP into a platform, or integration of the platform IP into a customer's SoC design flow. In the table we list the integration requirement in the left-hand column. Following that is an indication of whether the requirement is an interface (e.g., API or signal/timing definition) or format (e.g., mechanism of description). Finally, in columns three and four, we define whether the PrimeXsys Platform product family currently exposes these integration requirements explicitly in a private (platform standard), and/or public manner (open standard).

Briefly, the eight platform standards are:

(i)   Bus – the protocol (signal definition and timing), as well as extensibility of the bus-hierarchy provided in the platform context.

(ii)   OS Configuration Layer – call-structure for upper middle-ware embedded software components. For example, call structure for peripheral drivers and API for hardware-agnostic software implementations (i.e., using memory-map description for the SoC).

(iii)   Energy Management – at several layers of integration hierarchy, from exposure of clocking-grids at the physical level to enable gated-clocking, to energy-consumption reporting and a management policy controlled by a dedicated power-control unit at the systems level.

(iv)   Memory Structures – description of memory-map and hierarchy to allow user configuration, as well as auto-generation of integration test.

(v)   Trace / Debug – at several layers of integration hierarchy, from core integration requirements (e.g., embedded trace managers, and cross triggers), to model API. Also includes structure for capturing of data.

(vi)   IP Packaging – both for: the delivery of a complex piece of IP including multiple levels of design hierarchy, models, and design scripts; and tool recognition of structure of a piece of IP (e.g., interface definitions).

(vii)   Model Interfaces – interfaces for protected design sign-off models, as well as co-validation and soft-prototype models.

(viii)   Verification IP – interfaces to allow an extendible SoC verification environment to be provided (i.e., synchronization of tests), and format for auto-generation of tests (i.e., actions and file-readers).

The (dev) superscript in the table below indicates interfaces in current development within ARM's platform-based design initiative, described further in Section 4.4. The asterisk (*) indications are areas of currently active standardization work. ARM is expecting to provide these as ARM-community open-standards during 2003. For IP Packaging, an XML description standard is being generated under an ARM-community model, whilst we are also actively engaged in the efforts of the VCT group of the VSIA. For model interfaces, an AMBA 2 Transfer-Level SystemC Interface for SystemC models is now provided under an AMBA license agreement, while a transaction-level interface standard for abstract models is currently in progress. For verification IP, there is active collaboration between several major verification environment suppliers towards a common verification methodology and verification-component synchronization interface.

Table 9-1. Common platform interfaces and formats

Description	Interface or Format	Open Standard	Platform Standard
Bus (AMBA)	Interface	Yes	Yes
OS Config	Interface	No	Yes
Energy Management	Interface	No	Yes^{dev}
Memory Structures	Format	No	Yes
Trace / Debug	Interface & Format	No	Yes^{dev}
IP Packaging	Format	No*	Yes
Model Interfaces	Interface	Yes*	Yes
Verification IP	Interface & Format	No*	Yes

The following sub-section describes in further detail the interface standards used and being developed for hardware and model communication: the AMBA protocol family. Further description of some of the platform-standards interfaces is provided in Sections 4 and 8.

3.4   Component communication infrastructure

Since 1994, ARM has invested in the generation of open bus protocols for enhanced IP integration. As SoC modeling becomes a more crucial part of platform delivery, this investment will extend to the model interfaces for support of hardware-dependent software development and application software development. We describe the AMBA protocol family, then the modeling interfaces above that which will need to be openly standardized.

3.4.1   AMBA protocol family

AMBA is an open-standard, on-chip bus specification that details a strategy for the interconnection and management of functional blocks that makes up a System-on-chip (SoC). It is widely recognized as the most commonly used on-chip bus standard [9] and therefore boasts a wide range of ARM and 3rd party support including: bus fabric IP; CPU and peripheral IP; validation, verification and system design tools.

The AMBA specification encompasses a number of protocols intended to satisfy a range of performance, connectivity and power requirements. The latest revision of AMBA (3.0) introduces the highest performance AXI, together with existing AHB and APB specifications.

The Advanced eXtensible Interface (AXI) has de-coupled address and data buses, enabling then to be optimized independently. It also supports out-of-order completion and multiple outstanding transactions, maximizing data throughput and minimizing latency. The AXI specification defines the bus interface and protocols only, enabling the end user to chose or design the interconnect to match their bandwidth, connectivity and power requirements. For example, if out-of-order transaction completion is not required, then this can be left out of the interconnect whilst leaving the peripherals unmodified. Each link in an AXI bus topology is considered a point-to-point master/slave connection which implies that the interconnect is readily extensible and can be optimized at various points in the SoC design flow.

As well as the basic data transfer protocol, the AXI protocol also includes optional extensions to cover DMA, interrupt and low-power operation signaling.

The Advanced High-performance Bus (AHB) was the highest performance bus in the AMBA family before AXI was developed. It is suitable for medium complexity and performance connectivity solutions and currently has the highest levels of 3rd party IP support.

AHB-Lite is a subset of the full AHB specification and is intended for use in designs where only a single master is used. This may be a simple single master system or a multi-layer AHB system where there is only one AHB master on a layer.

The Advanced Peripheral Bus (APB) is designed for ancillary or general-purpose, register based peripherals such as timers, interrupt controllers, UARTs, I/O ports, etc. This is generally connected to the system bus via a bridge, which helps reduce system power consumption. The APB is very easy to interface to, with little logic involved and few corner-cases to validate.

3.4.2 Interface abstraction hierarchy

Platforms are complex pieces of design IP that must be provided with a set of soft-prototype versions. The reason that soft-prototype modeling of a platform is necessary is two-fold: (i) provide the correct level of design detail for the task performed against the model, and (ii) provide sufficient simulation speed for the design task to be performed. There are four basic tasks (use-models) that must be supported by soft-prototype models of a platform: (i) top-level architectural definition and embedded software development (supports: Platform Layer 3), (ii) micro-architectural exploration and middle-ware development (supports: Platform Layer 2), (iii) system, or micro-architectural, validation (supports: Platform Layer 2), and (iv) component implementation and integration/verification (supports: Platform Layer 1). Each soft-prototype provided with a platform must support a standard interface for user extendibility of the model.

Four levels of system abstraction are defined to cover the use-models we have just described. In Figure 9-4 below the system-level design layers are depicted against the basic SoC development flow, as first introduced in Section 2. The set of system-level abstractions that we use share many similarities with those described in a technical submission to OSCI [10], and

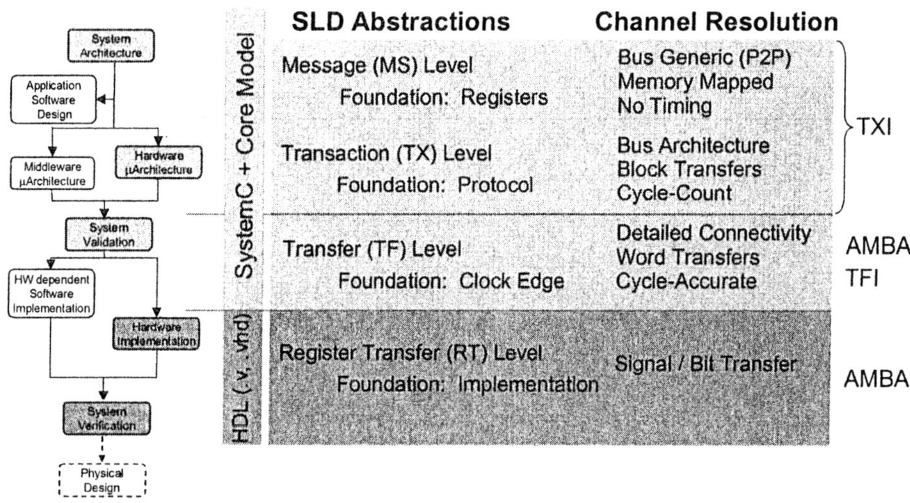

Figure 9-4. System-Design Flow against Platform Abstraction Levels

In Figure 9-4, the RT level of abstraction for the platform design is depicted in dark-gray, and constitutes the hardware implementation level. Above the RT level, the (loosely termed) 'transaction levels' of soft-prototyping are shown. At these levels of abstraction, we envisage that model extensibility will be provided through implementation of interfaces into a common open standard language, primarily SystemC. The platform models themselves may be provided in a separate modeling-specific format (not SystemC) that is more efficient for internal-core simulation, but which can be readily interfaced to the SystemC simulation kernel.

In the abstract system-modeling environment, three abstractions must be supported:

The (TF) Transfer Level – a cycle accurate translation from register-transfer level to transfer-level transactions. The simulation speed of the communication is gained through the efficient handling of abstract types and assembling of atomic (non-interruptible) action sequences into address/data transfers. This abstraction uses clock-based execution semantics, and is directly mappable into RT signals. For cycle accuracy, both blocking and non-blocking interface semantics must be supported. This modeling abstraction will execute in the 100 kCPS (cycles-per-second) range, sufficient for system validation.

The (TX) Transaction Level – a cycle-count accurate model of the system. That is, a datum or block-data request is completed (returned data or time-out/error) in single transactions, and time is indicated as 'time-passed' rather than events-per-clock-tick. Unlike at the transfer-level, there can exist models of the SoC in which only blocking communication actions are performed. The models of the bus and interfaces are sufficiently accurate as to be characterized by bus protocol and bus-hierarchy (i.e., domains, layering within a matrix, bridges, etc.). This modeling abstraction will execute in the 1 to 5 MCPS to provide for middle-ware development and micro-architectural exploration, such as bus and memory-management configurations.

The (MS) Message Level – a register-accurate level, with no or very abstract (e.g., "function call took y-time to execute") timing provided. There is no channel modeling between the components at this level as point-to-point communication is used, with the exception of shared memory spaces. This modeling abstraction is sufficiently accurate for embedded software development as it exposes the OS configuration layer and a register-level, or programmers-model, description of the SoC. This modeling abstraction will execute in the 10 to 100 MCPS.

For each level of abstraction, a system-level interface must be defined. These interfaces should be open standards that are expressed in a comprehensive manner to describe not only the communication API, but also the mechanism of refinement from one abstraction to the next [12]. The concept of interface hierarchy is depicted in Figure 9-5 with the system-abstraction levels defined on the left-hand side of the figure. This figure shows that, while there may be many specific protocols at the RT level, these can be reduced into a smaller set of configurable interfaces as the design is abstracted. At the top-level, the system model becomes bus-generic, though this generic transport-layer can be configured to express the 'protocol personality' of a bus. A protocol personality will include, for example, length of burst support, details of request/grant procedures, error-handling, side-band information, etc.

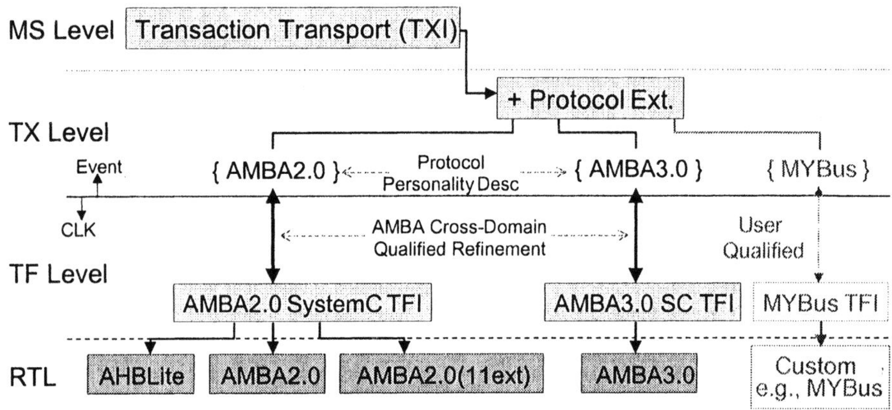

Figure 9-5. System-Level Interface Hierarchy

Besides the interface hierarchy, the most important point expressed in Figure 9-5 is the delineation between the system abstraction levels. There are two significant boundaries, one between the RT level and TF level that is a language boundary (i.e., multi-level simulation will imply multi-language) but otherwise a straightforward translation; and one between the TF and TX levels, a communication mechanism boundary. The distinction between the TF and TX levels of abstraction is a fundamental mechanism of model-execution: a clock-based execution semantic is used at Levels RT and TF, and a synchronous-reactive execution semantic is used in Levels TX and MS. Linking or refining models across the TF/TX boundary is not intuitive due to the lack of an atomic-unit of time in the TX/MS abstractions, unlike the clock-period for simulation advancement in the TF/TX abstractions.

Although a separation between the event-based TX and MS levels also exists, this difference can be characterized within the 'protocol personality description' that enables transfers to be weighted with an appropriate delay. There is neither a difference in execution semantics nor language in moving across the TX/MS abstraction boundary.

Models of ARM platforms are provided conforming to the interface hierarchy described above. ARM is currently providing SystemC extensibility to its Instruction Set Simulator models that enable integration into SoC environments supporting MS and TX abstractions. We also provide as an open-standard the AMBA 2 SystemC Transfer-Level Interface Specification that defines requirements for building of AMBA interface class-libraries (ACLs) in SystemC. With partner companies, ARM is actively investigating support for qualified translation from TX level models to AMBA 2 and 3 interfaces at the TF levels and below. Application of the modeling abstractions to the platform-design process are detailed in Section 5 and Section 6 of this chapter.

4. DESIGN SUPPORT: HARDWARE INTEGRATION

4.1 Introduction

Hardware integration is eased by the provision of standardized interfaces, as described in Section 3.3. However, standard interfaces are not sufficient for providing an efficient hardware platform integration strategy. Platforms must be able to be provided interface-protocol checkers (for validating compatible IP design such as the AMBA Compliance Test-bench (ACT)), integration examples (such as the AMBA Design Kit for bus architectures), and SoC resource allocation tools (such as PMAP for memory allocation). This section describes how ARM addresses the most critical hardware integration issues for soft platforms. In this section, we describe the platform-based design support provided for hardware integration, Platform Layer 1. Hardware integration and verification is supported by the clock-based RT and TF system model abstraction levels.

4.2 Bus-generation and hardware integration

To enable customer integration of hardware components, ARM provides two products: the AMBA Design Kit (ADK), and the AMBA Compliance Test-bench (ACT).

The ADK provides a generic, stand-alone environment to enable the rapid creation of AMBA-based components and SoC designs. Containing a rich set of basic components, example system designs, example integration software and synthesis scripts, the ADK provides the common foundations for product design based upon the AMBA interface. This frees up engineers to focus on application-specific issues and "value-add" components.

The ADK is comprised of over 50 fully validated components, which can be broadly categorized as interconnect (e.g. decoders, arbiters, bridges), reusable peripherals (e.g. interrupt controller, timers, watchdog), example peripherals (master and slave), ARM CPU wrappers and test components (BFM, memory models etc). All are available both as VHDL or Verilog HDL code.

The ADK introduces the concept of "black-box" interconnect, where an AMBA interconnect can be auto-generated based on a number of parameters and plug in arbitration and decode schemes. This is then presented to the system integrator as a black-box of master and slave ports to which can be attached peripherals or bridges to other interconnect.

ARM's ACT enables IP component developers to demonstrate that its AMBA interface is fully compliant with the specification [13] and can therefore be seamlessly integrated with other compliant IP. An IP component is granted AMBA compliance when it has been observed to experience all of a predefined list of protocol scenarios (coverage points) without breaking any protocol rules. ACT is integrated into the HDL simulation to provide this capability of AMBA protocol checking and coverage analysis. The protocol and coverage checkers gather the evidence required to grant AMBA compliance to the device under test (DUT).

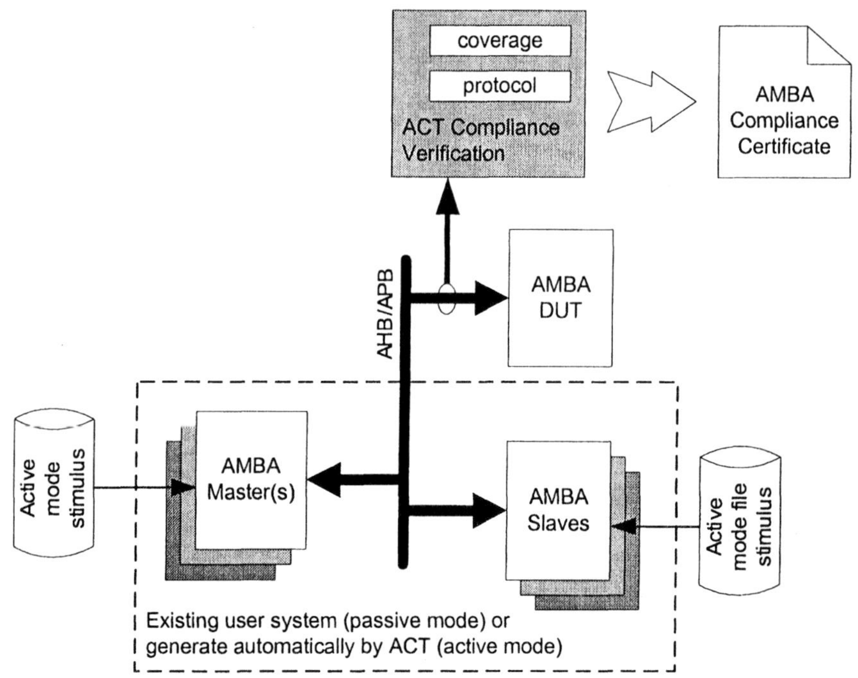

Figure 9-6. Architecture of the ARM ACT product

Figure 9-6 shows a generic ACT structure. When configured in 'active-mode', ACT generates a testbench for the DUT including stimulus driven bus functional models of master, slave and arbitration components. In passive mode the protocol and coverage checkers are overlaid onto an existing user testbench.

4.3   Configuring the Platform

In addition to the AMBA interconnect generator described in the previous section, the platform must support a capability to automatically generate and manage the complete address, DMA and interrupt maps. This requires provision of template-based RTL and models for the address-decoder, interrupt-controller and DMA blocks. Significant co-ordination needs to be achieved across platform components and deliverables by creating shared schemes for describing component parameterization, and providing methods for generating the configured RTL/models and verification vectors. ARM's own production version today is known as PMAP.

PMAP, or 'peripheral map', defines the local register descriptions for each peripheral, including complete register bit maps and reset values. These are concatenated together by a top-level script-file that defines the platform or SoC memory map[9]. From the top-level script, the RTL for the top-level address decoder is generated along with integration test vectors for the SoC. The integration test vectors can take the form of either instructions for an AMBA File-Reader Bus Master (FRBM), or C-code for compilation onto a core.

In addition to automatic generation, a PMAP description for an individual peripheral can be optionally extended with a set of canned vectors to perform self-test or run initialization code. This allows basic software integration to be carried out in a simple simulation environment and is particularly useful for toggling interrupt lines to verify the correct integration to the interrupt controller.

PMAP is currently based on an in-house format and script, but it can be described with an XML-based schema. This open-format approach will encourage support of platform configuration in a standard way across a variety of EDA tools

4.4   Extending platform control structures

As described in Section 3.3, full platform integration requires definition and standardization of interfaces to more than just the bus infrastructure and memory configuration. In particular, significant development activity within

---

[9] It is not generally desirable to change the base address map of the platform as such a change may necessitate kernel re-write. This problem arises with OS ports when key peripherals are tightly coupled to the behavior of the kernel. It is, however, generally feasible to extend the address map to hook in communications-oriented peripherals.

ARM is being applied to two interfaces not today shipped with platform product: the energy-management interface, and the trace-debug interface.

System-level energy management is a complex task that involves communication between an application or operating system and the hardware platform. This requires a hardware-dependent software abstraction (Platform Layer 2) of the energy management system that is extendible when new components are integrated. At the hardware level (Platform Layer 1), a scheme and protocol for communicating power status from a central energy management controller to peripheral is being defined. This will permit definition of, and plug and play with, different power domains on a configured platform. A first step in this development is the collaboration between ARM and National Semiconductor to define the off chip interfaces between a power management controller and voltage control IC to effect dynamic voltage scaling.

System debug support for a platform requires both hardware and software support. ARM pioneered the use of in-circuit trace and emulation (ICE) and it is natural to extend this to the complete platform and peripherals. From a hardware perspective this requires the definition of a standard multi-master ICE and trace protocol. Trace is already being extended within ARM to provide real-time trace capability for an AMBA bus. For multiprocessor support, a trace/debug hardware construct must also define a cross-triggering standard.

The software support for system-debug is provided by ARM's Real View Debugger product, a product that already supports a multi-master ICE and trace protocol. The Real View Debugger product allows a complete description of a new peripheral to be rapidly integrated by providing the software developer with symbolic awareness of the platform and its extensions. The Real View framework also permits the straightforward integration of source level debuggers for DSP or other processors into a heterogeneous SoC environment.

5.   DESIGN SUPPORT: HARDWARE DEPENDENT SOFTWARE

5.1   Introduction

The point at which hardware-dependent software, often termed firmware, and the hardware itself comes together is often the first time a system exists in its whole form. Traditionally, this has been done using hardware prototypes where errors in the system design have been corrected with software workarounds or, much more costly, using re-spins of the hardware. New techniques bring hardware and software together earlier in the design cycle and seek to eliminate the need for the costly hardware re-spins [14]. In this section, we define the platform-based design support for Platform Layer 2. This encompasses modeling abstractions for the clock-based TF system abstraction level and the TX cycle-count abstraction, as well as the process of model refinement. This section also includes a discussion on hardware prototyping support for hardware dependent software development.

5.2 Hardware/software co-verification

Early hardware/software co-verification circa 1995 [15] introduced embedded software to hardware simulation using retargeted code and memory integration libraries in which accesses to hardware memory were replaced by accesses to a Bus-Functional Model (BFM) simulating a processor bus interface. Instruction Set Simulators soon replaced the retargeted embedded software creating a better-integrated tool-chain. Most recently, high-speed cycle-accurate models have been introduced and provide an accurate, but efficient integration of hardware and software. Additionally, co-verification kernels have introduced sophisticated optimizations that allow designers to trade off bus cycle-accuracy for speed.

Whilst co-verification remains an efficient way to interface embedded software to an RTL representation of a design, the software is still introduced to the hardware relatively late in the design of the complete system when the RTL implementation is well under way. Co-verification is also an expensive technology that is not popular with software developers who are unfamiliar with its hardware-centric view of the system.

5.3 Hardware prototyping

As before with application software development, hardware prototyping using field-programmable gate arrays (FPGAs) or FPGA-based multi-board prototyping tools introduces software to real hardware and executes that software on a platform capable of multi-megahertz speeds. Integrated software tool-chains ensure that the software developer has a consistent view of the software running on the target regardless of whether he or she is using a software emulator, a final product or the hardware prototyping tools.

For hardware-dependent software, however, FPGA prototypes can provide too little visibility into the description of the hardware. For the system architect, they are not well suited to rapidly exploring new configurations of system hardware in the context of software benchmarking. A system designer may, for example, wish to explore how different bus or cache architectures affect the performance of an operating system just as much as they wish to tune an operating system to a particular hardware micro-architecture. This is more directly achievable with soft-prototyping approaches, particularly transaction-level modeling.

Although hardware visibility issues exist with hardware prototyping that make it less desirable for middleware development, the execution speed of the hardware prototype is very applicable to application software development, as described in Section 6.2.2.

5.4   Transaction level modeling

Transaction-Level Modeling (TLM), as introduced in Section 4.2, seeks to bridge the gap in the current available hardware/software integration methodologies. By raising the level of abstraction of the hardware described for the system, designers can produce more efficient system simulations. Speeds of circa 100kHz for transfer-level simulation and 1MHz for transaction-level simulation are achievable using the abstraction technique in a suitable modeling language such as SystemC.

Transfer-level (TF) modeling provides a cycle-by-cycle mapping from one bus transfer cycle to its representation in a complete bus protocol. Speed increases are achieved by a combination of high-value IP models written to simulate efficiently in such environments and the reduction of a number of signal events in a protocol-cycle to a single-cycle operation. At the more abstract transaction (TX) level of design, an entire block-data transfer such as an ATM packet or an AMBA AHB burst can be represented by a single transaction. The timing of these transactions will always be known to within a reasonable approximation because the protocol is well understood. Each transaction can be converted to cycle-based timing using an adapter.

5.4.1   TLM example system

To understand the architecture of a typical TLM system, a simple example is shown in Figure 9-7 in which four IP components: 2 masters and 2 slaves are connected through a shared communications fabric supporting the AHB Multi-Layer bus protocol. This example applies to the clocked system-abstraction levels, transfer (TF) level and implementation (RT) level. The architecture exists within a single clock domain. Masters, slaves and the hierarchical channel model are all clocked components. Arbiters and decoders, as part of the bus fabric, are reactive, un-clocked models.

The communications fabric is modeled by three components: a hierarchical channel model that manages state of component connectivity, an arbiter that receives requests and allocates the shared channel based upon the priority of the requester (master), and a decoder that resolves addressing into specific block connections (transfers from/to). The arbiter and decoder, though distributed in hardware, are expected to be modeled through use of monolithic SystemC blocks (single interface to each).

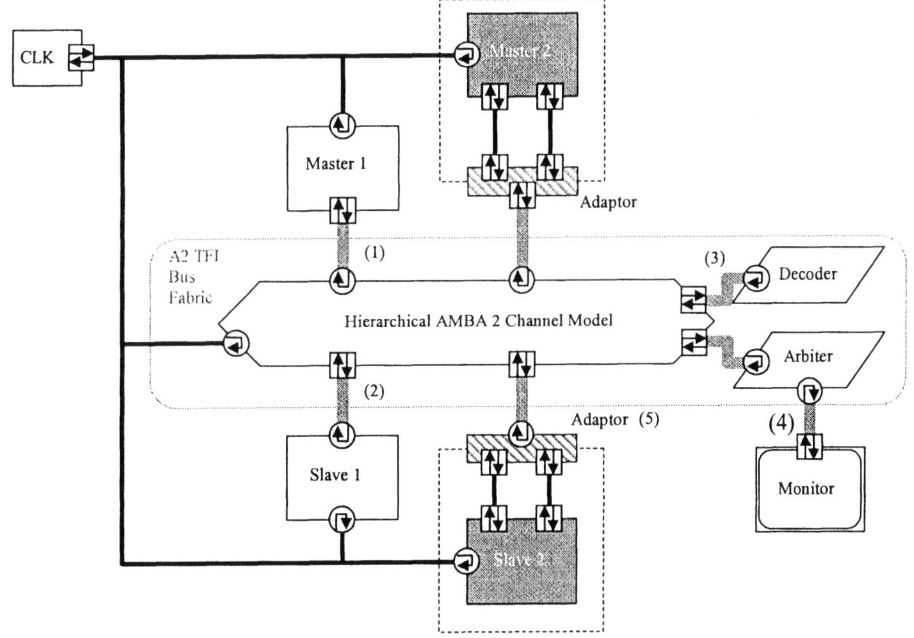

Figure 9-7. IP Models and an AMBA bus fabric in SystemC

Masters in such a system might be processor models or memory controllers, whilst slaves might be peripherals such as a UART. Masters are clocked models that create data structures representing bus transactions, masters pass these structures to the bus fabric. Slaves are clocked models that are invoked by the bus fabric and are given the bus transaction data structures to operate upon.

A critical part of this methodology is the ability to migrate to RTL design and retain a verification infrastructure for the complete system. The verification methodology is described in detail later (Section 7), but the introduction of RTL models is achieved using adapters to the transfer interfaces of the SystemC models. Since there is a known mapping between AHB transactions in the SystemC world and the signals of a physical AHB, such adapters can be generic and guarantee the correct operation of the device.

A single monitor (e.g., waveform analysis tool) is shown in the example architecture, connecting through a monitoring interface to the arbitration component. As the arbiter maintains a list of channel requests and masters the allocation of the shared channel, this component maintains specific information about the current and pending state of system connectivity not maintained elsewhere. Simple transfer monitors may also be placed on master / slave interfaces, but these are not depicted here.

6. DESIGN SUPPORT: APPLICATION SOFTWARE

6.1 Introduction

This section describes the platform-based design support for embedded software development, Platform Layer 3. The section starts with an overview of current approaches for software development on custom SoC [16] then describes the more efficient and cost-effective support possible in platform-based design. We describe both the modeling support for Platform Layer 3 encompassing the message (MS) level of modeling-abstraction, and the hard-prototype support also critical for application software development.

For context on the delivery of platform models, we utilize the definition of IP Creator, IP Licensee, and IP User provided in Section 2.

6.2 General application software environments

Most application development environments do not adequately consider the requirements of the IP deployment strategy. Typically in the case of the three contributors to a completed platform-IP based product, the IP User will be responsible for application software development, but will do so based on an operating system port developed primarily by the IP Creator. Since there is no requirement for a business relationship between the IP Creator and the IP User, there must be sophisticated, well-supported development tools in place to support new platforms.

6.2.1 Host-retargeting

Many OS vendors provide an environment in which the OS itself is simulated using a native port of the OS (such as the Symbian WINS emulator [7]) and the application code recompiled for the host platform. Whilst this approach results in excellent software execution performance, often approaching the speed of the host platform itself, there are a number of drawbacks in this mode of application software development.

Companies employing host-retargeting will need to tool up their application developers with host compilers as well as target compilers. The resultant code is also not guaranteed to work due to the subtleties of the different representations of the OS on the target and the host. Threading models available to user processes on the host and to a kernel on the target will differ greatly and may affect the execution of applications such that correct execution on the host does not guarantee execution on the target.

In the case of complex platforms, representations of the underlying hardware must be provided to the user to ensure correct function of an application written for a specific device. Even simple device geometry issues such as the size of the screen on the host and the target can affect the application significantly. Some OS vendors have solved this problem by introducing a virtual machine that traps hardware accesses and emulates the underlying hardware using models. This added complexity typically reduces the performance of the emulation significantly, thus negating much of the benefit.

6.2.2 Hardware prototyping

Prototyping of systems using FPGAs and FPGA-based multi-board prototyping tools is becoming increasingly popular as the complexity of hardware and software increase. These tools allow high-speed execution of software on real hardware, often allowing the interfacing of physical devices such as networks that provide true physical data. In the past a drawback of FPGA prototyping for the IP Creator has been the protection of their IP. Recent developments in the area of secure FPGAs has allowed IP creators to offer pre-compiled FPGAs containing the pre-configured platform IP.

Using hardware debug and trace tools, software developers can gain excellent debug visibility into the software running on a target. Debug tools need to be aware of the underlying OS and provide information about OS-level primitives such as threads and semaphores if they are to provide sufficient debug infrastructure for an application software developer looking to debug complex OS-dependent software.

FPGA-prototypes do suffer from the complexity of new platforms and large systems and whilst newer and larger FPGA devices are constantly available, complete systems will always span a number of FPGAs. This is particularly true in the case of the complex IP deployment chain described in Section 2 as each party in the chain will seek to encapsulate their IP in their own secure FPGAs. When this proliferation of FPGA devices occurs, the performance of the resultant system decreases and falls below that required by application software developers.

Most significantly for the devices that are intended to be open to a huge range of application developers such as set-top boxes, PDAs and new cell-phones, hardware prototypes are expensive. If a product supplier is to encourage the development of software for their device, the development environment must be inexpensive or even free.

6.2.3   Instruction-Set Simulation

Instruction-Set Simulators (ISSs) are typically available for all programmable devices and often provide a low-cost route to the emulation of target software. Modern ISSs also typically include simple infrastructure for modeling the programmer's interfaces of peripheral devices so that the execution of hardware-dependent software such as an OS is supported. They operate at the MS-level of system abstraction. The models are sufficiently accurate, and sufficiently fast in execution, that application software can run on top of the OS and the application developer can debug the application using the same debug and trace tools available for a live target such FPGA prototyping.

Execution speeds of ISSs are constantly improving and the latest generation of software emulators can execute software in the performance range of some FPGA prototypes at around 10 MCPS. Since they are software products, they can be distributed inexpensively as part of a software developers' toolkit. The latest embedded products, however, require higher speed processors and the performance of even a 10 MCPS ISS is unlikely to be suitable for application development. When models of the hardware subsystem are included, execution speeds are affected as much as they are in the case of the host-retargeting OS environment.

6.3   Platform software development environments

ARM currently offers platform solutions in all three areas described above for application software development and each of these solutions provides a unique solution to the software development problem. What is not addressed by any of these environments is how to offer an execution environment in which host-targeted software can be represented at a low cost but at speeds that are suitable for application software development. Specifically:
–   Host retargeting does not give the developer the confidence that is gained
    by targeting software for the physical hardware target and using the
    target OS.
–   FPGA-prototyping does not provide a low-cost environment for the mass
    proliferation of a platform to thousands of developers.

- ISSs are not fast enough to execute the complex application software and OS kernels that are present in modern embedded devices.

To address the low-cost requirement of the application software development environment, a software solution is best as the costs of replication are low. The traditional software emulation provided by ISSs is insufficient in terms of performance for application software, so new emulation techniques are required. The virtual platform must include models of the hardware represented by the device and these must be presented to the developer in such as way as to emulate the physical design of the target device. In the case of a cell-phone, this might be in the form of an image or "skin" of the device that is updated by the software emulator in real time.

In terms of where such a product fits into the development cycle of a device, this does not fall into a simple top-down approach. Instead, it is likely that the high-speed platform emulator will be developed once the product is nearing the end of its development and is well understood. The IP Creator will initiate the deployment of the base platform for the emulator and encourage the IP Licensee and IP User to extend the emulator to match particular devices running specific OS ports. The goal of this deployment chain is to ensure that the platform emulator is available at the time of the launch of the product such that it can be supplied to application developers to encourage them to develop new software for the new device.

7. QUALIFICATION: PLATFORM VERIFICATION AND VALIDATION

7.1 Introduction

Although reuse methodologies in design implementation have become an essential part of the design approach for complex SoC development, reuse of the verification environment is much less common. Traditional approaches to verification do not enable reuse of the verification effort from project to project. Typically test stimulus is still handcrafted, from scratch, for each new SoC project using a mix of C, RTL and assembly language. Acceleration and emulation hardware, and new verification tools and languages offer enhancements to the conventional flow, but they do not on their own improve verification IP reuse.

Due to its architectural stability, a platform architecture is an excellent vehicle for the provision of an extensible SoC validation environment. Today, platform verification and validation environments are being supplied by ARM that explicitly support coverage-monitoring and test-bench synchronisation on a complex SoC. In this section we present a validation flow that directly addresses these problems. This flow has been industrially proven with the release of the ARM PrimeXsys Platform in Q1-2002 [1]. The deliverables in this ARM platform product are described in Section 8.

7.2   Extensible verification and validation

The section below describes the technical structure for support of a verification and validation (V&V) methodology that specifically emphasizes reuse, extendibility, and configurability [17,18]. The section following this describes the process for platform V&V using this structure. Verification in this context refers to functional verification. Validation in this context refers to checking the implementation against the original design intent.

7.2.1   XVCs: reusable V&V components

This methodology is founded on the definition and use of 'extensible verification components' (XVCs), which are reusable components that can represent either a SoC component, or an external device interfacing to a SoC component. An XVC is typically implemented in an HVL (high-level verification language, such as 'e', Vera or the SystemC verification library), and can be used to create stimulus or monitor activity, and contains device-specific/designer knowledge. XVCs contain a library of test operations from simple I/O to executing stimulus files; these are called 'actions'.

Although current implementations of actions contain pseudo-random and/or directed test stimulus, and temporal checks for protocol, etc., future implementations may also contain assertion-based verification concepts that are used to capture designer knowledge about a device under verification (DUV).

An external test scenario manager (XTSM) is used in a 1:n connection to XVCs to schedule and synchronize selected lists of actions. The XTSM is programmed via a simple input file, which contains a group of XVC action sequence calls. This input file is designed to represent a typical system usage scenario. Each XVC connects to the XTSM via a two-way messaging API, and new XVCs are added to the test environment simply by registering with the XTSM at runtime.

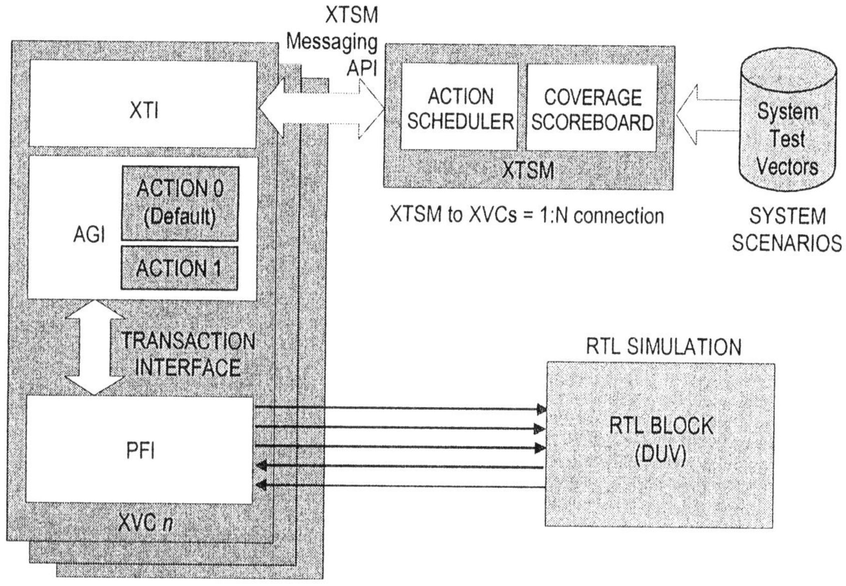

Figure 9-8. XTSM/XVC Structure

The external test interface (XTI) receives command actions from the XTSM. The XTI communicates with the action generator interface (AGI) and synchronizes the execution of action sequences. An action can be any number of test sequences to be applied to the DUV. The user can add any number of test actions to the AGI and this process gives the XVC its extensibility. Finally the AGI sends transaction requests to the peripheral facing interface (PFI) that in turn converts the transactions into signal driving and sample operations. Sampled data flows back through the PFI into the AGI, which in turn communicates status information back to the XTSM via its XTI. The PFI can be viewed as a hardware abstraction layer for the XVC.

The XVC architecture explicitly supports multi-layer V&V consistency. The same test-suite may be used at the hardware-dependent software stage with, for example, a SystemC transfer-level model of the platform, and at the hardware integration level with, for example, Verilog IP. If the transactions are carefully authored, this can be extended to the software-development layer. This multi-layer approach is enabled by using a transaction-based encapsulation of actions, and using the PFI to do the abstraction translation.

7.2.2   Coverage monitoring

One of the uses of the XTSM is to also co-ordinate functional coverage data collection and score-boarding. Having a central store for coverage data also brings the advantage that this data can be read 'on-the-fly' by individual XVCs. This important feature allows some degree of automated testing in that random stimulus data can be generated and constrained using feedback from the database managed by the XTSM. Techniques using this or similar approaches are often termed as 'directed random testing'. Systems can be driven until all coverage goals are met, or report if a particular run did not hit specific functional coverage goals.

7.2.3   System-level platform verification

Performing system-level verification should ensure that the individual blocks have been connected correctly and that they perform as intended, in context. Three testbenches are used as part of the system-level verification solution. These testbenches are optimized for:
- **Hardware Integration (Platform Layer 1):** Checking of hardware DMA, IRQ and bus interconnect;
- **Software/hardware Integration (Platform Layer 2):** Checking that software can boot and initialize the hardware through driver routines;
- **System Validation (Platform Layer 2):** Checking the performance and latency within the system, along with corner case scenarios.

Each of these stages of SoC platform verification is explained below. A platform environment for testing of the application software, Platform Layer 3, is provided separate from this SoC verification flow. Software test for Platform Layer 3 is enabled through provision of hard and soft prototyping environments for the platform (see Section 6) that can be targeted through a common debugging front-end.

7.2.3.1   Hardware integration testbench
The primary objective of this testbench is to verify that the hardware blocks that comprise the system have been connected together correctly. This testbench is not designed to collect functional coverage or to verify software driver integration.

The system integration testbench enables rapid verification and debug of the block interconnectivity. A typical scenario would be to identify incorrectly connected control and/or address signals, IRQ or DMA request lines. A CPU model is deliberately omitted from this hardware integration testbench. This ensures the fastest possible turnaround, since no software is being run on the CPU. In place of test software, instructions are injected from a simple text file into the appropriate CPU 'socket' within the design. A file-reader bus master model (FRBM) is used to emulate CPU bus reads/writes. The instruction file itself is automatically generated from a set of easily modifiable configuration templates that describe the locations and capabilities of the system peripherals. Combining this with a memory-map configuration file to control the number of tests generated provides an instruction test sequence that reflects the system connectivity. Note that no XVCs are required in this testbench environment, as simple RTL blocks are used for 'loop-back' connections on any external device ports.

7.2.3.2   Software/hardware integration testbench

The primary objective of this testbench is to verify that software can successfully boot and interact with the system. This testbench is not designed to collect functional coverage, or to verify block interconnect at signal level. The architecture of this testbench is shown in Figure 9-9.

Primarily for performance reasons, a C-based Design Sign-Off Model (DSM) of the target CPU is used. As the name suggests, this is a detailed model of the processor, capable of running software to meet the sign-off requirements of the design. Typical usage of this testbench would be for booting the system, configuring and accessing the memory sub-system, running a device driver through the CPU to exercise system peripherals, and stimulating data block transfers through the system.

A user extendible C++ test manager package supports the software test code environment. This manager allows the verification engineer to focus on developing effective test sequences, without the overhead of integrating the tests together. A library of generic (non-OS specific) peripheral device drivers forms part of the standard test suite.

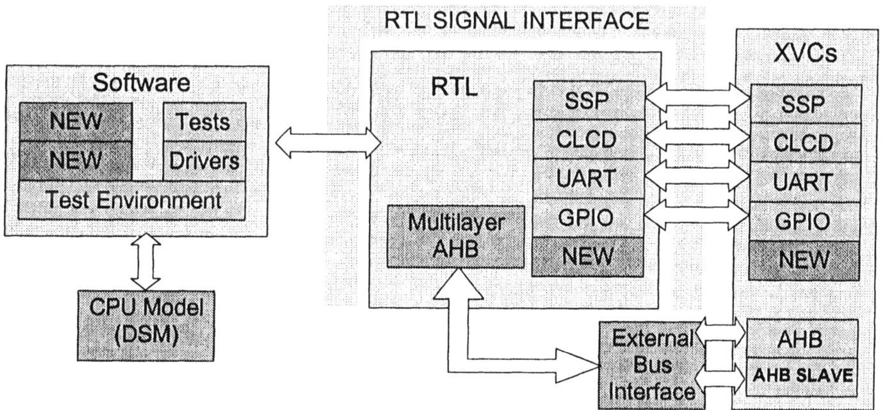

Figure 9-9. Software/hardware integration testbench using a CPU model and XVCs

7.2.3.3   System validation testbench

The primary objective of this testbench is to allow efficient generation and monitoring of complex hardware block interactions or corner-cases, and at the same time measure functional coverage and system performance. This testbench is not designed to verify software driver integration or to verify block interconnect at signal level.

Typical scenarios might include the bus master programming a peripheral, external stimulus arriving at that peripheral, and another peripheral simultaneously generating an interrupt request to be serviced. In this testbench the CPU DSM is not used.

To ensure that the system validation process attains good coverage, many scenarios may be sequenced, potentially creating a significant simulation burden. Removing the need for the CPU DSM improves the efficiency of the testbench. In its place an XVC Bus Master is used, which combines the ability to read from files, and drive the bus with a wider programming capability. For convenience, this master can also read and drive stimulus used by the FRBM from the Integration Testbench.

In contrast to the CPU in the software/hardware integration testbench, an XTSM is used to drive the testing and to provide sequencing and scheduling of the XVCs – for example, to sequence actions for a Master XVC and for it to program system blocks to output data. Corresponding XVCs respond to output data using actions that are also specified by the XTSM. This environment allows the creation of complex test sequences to be abstracted to a very high level. This testbench environment is easy to control and extend, and the verification team can customize an off-the-shelf XVC providing for reuse of the entire validation environment. The adaptation of the testbench architecture to include a test-scenario manager and mastering XVC as replacement to a core model is shown in Figure 9-10.

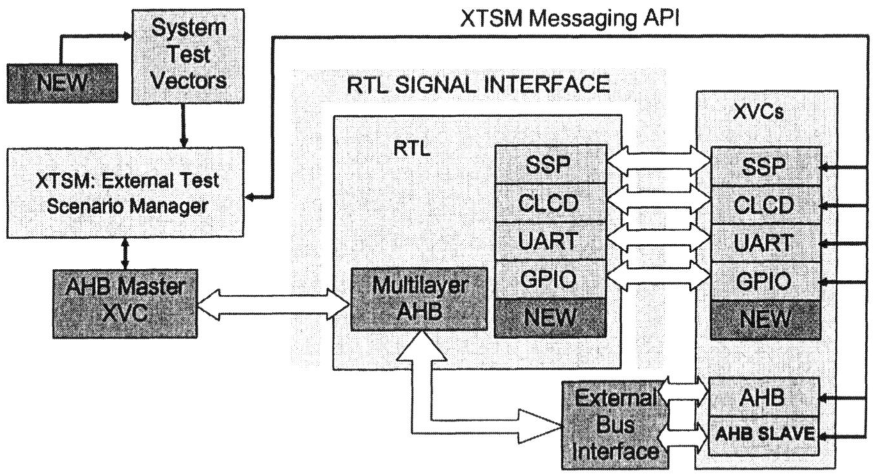

Figure 9-10. System validation using XTSM/XVCs to generate 'typical usage' scenarios

8. WORKED EXAMPLE: PRIMEXSYS PLATFORM

ARM is developing and delivering the PrimeXsys Platform (PXP) Product-family of systems IP components. Each member of this product family adheres to the principles outlined in the preceding sections. In addition to the hardware and software IP-bundle that has been optimized and validated to support the demanding requirements of consumer OS and embedded RTOS, the PXP products each contain a range of deliverables to improve the time-to-integration of a customer's differentiated SoC design. This section describes the ARM1136JF-S^{TM} PrimeXsys Platform product that became available to customers in the first quarter of 2003. An earlier generation of this platform, the ARM926EJ-S^{TM} PrimeXsys Platform, has already been shipping to customers as of the first quarter of 2002.

8.1 Establishing a PrimeXsys Community

PrimeXsys Platform products are founded on support of the following guiding principles: (i) a fully validated integration of a software and hardware subsystem with ported OS; (ii) a well-defined set of standards-based extension ports to allow easy integration of differentiating IP blocks on to the base platform; and (iii) an extensive third-party program consisting of IP components and supporting EDA tool infrastructure

The final of these three points, the business environment for platform support, must be emphasized. A platform provider must ensure that standards for extensibility are backed by a widespread adoption program to ensure that a broad base of 3rd-party IP is available for integration, both for hardware as well as for software and OS support. ARM has established this with the formation of the PrimeXsys Community Program. The PrimsXsys Community Program is founded on three types of corporate partnership: (i) component providers of enabling IP (hardware and software) conforming to the PrimeXsys Platform extendibility standards as introduced in Section 3, such as the AMBA 2.0 compliant peripherals of the Synopsys® DesignWare™ library and the OS partners who include the providers of: Windows CE, .NET, Symbian OS 7.0 and VxWorks; (ii) EDA and software-development tool partners who ensure that design environments explicitly support the platform standards, such as AMBA-compliant support of SoC-generation in the CoWare® N2C Design™ System tool, verification in the Verisity® Specman Elite™ environment, and support of emulation with the Aptix® System Explorer™; and (iii) software developers who create application packages for driving the platform through a hardware-abstraction layer provided by the operating system port. This broad technical community that is coordinated around the PrimeXsys extensibility standards is the key business driver for adoption of ARM platforms.

8.2   Architecture of the ARM1136 PXP

The ARM1136 PXP is based around ARMs latest generation ARM11 synthesizable processor, the ARM1136, designed to support demanding next generation applications. The processor supports the latest v6 instruction set, which is targeted to improve OS and media application performance. The ARM1136 processor features a complete level-1 memory system including a memory-management unit coupled with an advanced cache architecture. To address the increasingly demanding debugging requirements of advanced real time software, the ARM1136 PXP integrates an embedded trace macrocell and buffer to complement the integrated ICE feature of the core. The fast interrupt response of the processor is supported through the integration of a vectored interrupt-controller in the platform. More details on the ARM1136 can be found in [19].

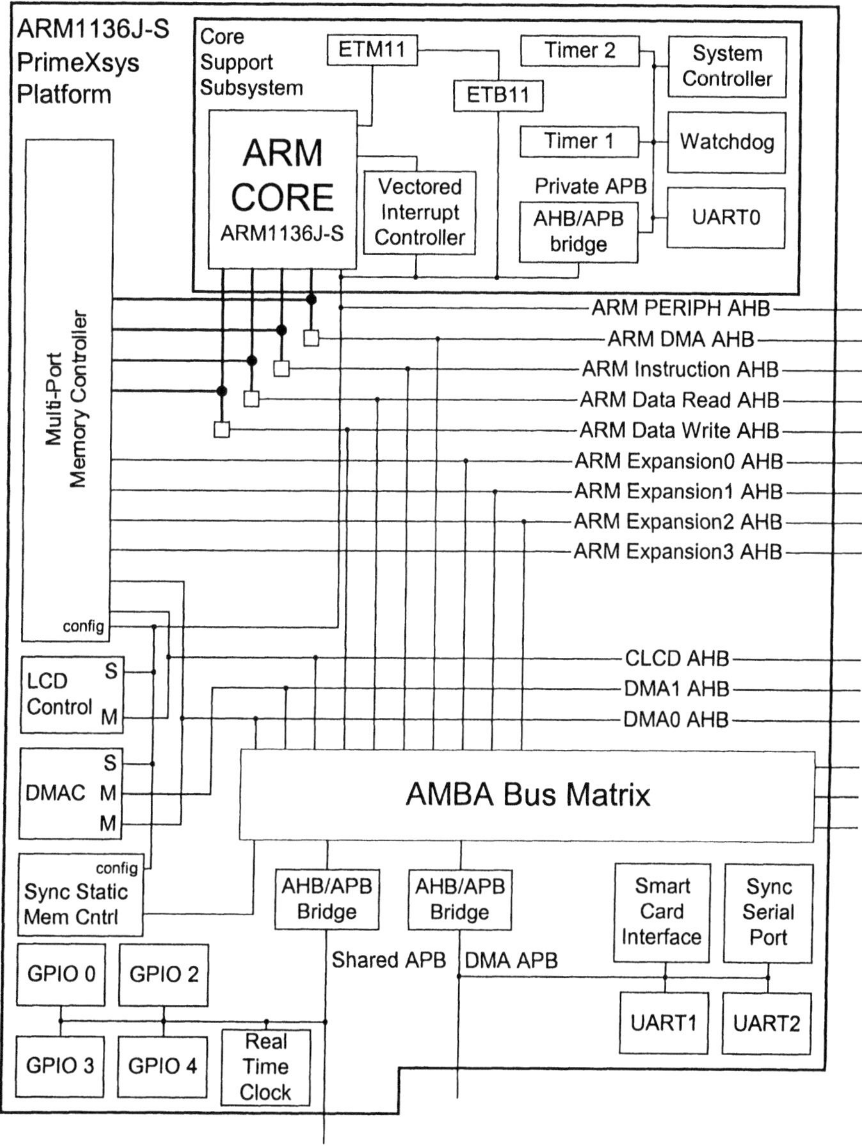

Figure 9-11. Architecture of the extendible ARM1136 PXP platform

The Platform architecture, illustrated in Figure 9-11, is based around a multi-layer AMBA bus matrix. This multi-layer structure uses nine full multi-access buses (MAB) bus segments and connects the CPU, peripherals and platform extension ports in a flexible manner to provide maximum concurrent bandwidth in support of the most demanding applications. To ensure smooth passage of information around the platform, the processor's own instruction, data and DMA channels are supported by a multi-channel DMA block, an advanced multi-port memory controller (MPMC) and static interface. Between these components multiple simultaneous data transactions are supported between platform components, bulk and static memory.

The bus matrix features several extension ports, which are described in more detail in Section 8.3. These allow additional peripherals or subsystems to be connected directly to one of the platform's bus segments, including the processor local busses. In addition support is provided to allow an external master to access all of the platform peripherals including the memory controller. This facilitates the implementation of unified-memory interface SoC systems.

The platform architecture is completed by two sets of peripherals that enable validated standard OS ports to build efficiently. The first set or peripherals comprises a real-time clock, watchdogs and timers, UART and GPIO. This set provides the basic hardware support required to boot and communicate with an operating system. The second set comprises a LCD controller, synchronous serial port and smart card interface. This set provides complete support for a GUI based OS platform. These peripheral sets provide the platform with a defined set of capabilities and therefore define a common Hardware Dependent Software (HdS) abstraction for the operating systems port. This is an important concept in the delivery of a configurable and extendible platform design since changes to the HdS layer would necessitate re-porting of the OS.[10]

The platform architecture also supports address re-mapping to dynamically re-locate the physical location of memory. This supports boot from static memory and subsequent reuse of the low memory area for interrupt vector code.

8.3   Support of hardware integration

The bus fabric features a number of extension ports to which additional hardware blocks or subsystems can be connected. This is a key example of how a standards-based framework enables easier integration. Each extension port of the platform conforms fully to the AMBA standard, so any peripheral-block compliant with this standard can be 'plugged' into the extension ports with a high degree of confidence. We provide the AMBA Compliance Testbench (ACT), (see Section 4.2), to verify that a peripheral-

---

[10] An active area of research within ARM is the pursuit of a standard way of expressing an extendible HdS platform abstraction for memory controllers. This will allow new memory variants to be quickly supported without affecting the platform configuration.

block interface has been tested for all possible AMBA bus transactions or events. The ACT provides AMBA compliance certification that a platform integrator may take as an unambiguous contract of integration quality when engaging with a 3rd-party IP provider.

Verification of 3rd-party IP blocks is also supported through the ARM Integrator product. This is an extensible FPGA platform that enables rapid prototyping of peripherals. The platform is provided as a core module into the Integrator environment that can then be extended to prototype the functionality of the configured SoC.

8.4   Support for software development

Software development is supported by two extendible views of the platform: the Integrator FPGA product, and an ARMulator platform soft-prototype.

Into the Integrator product, the ARM1136 PXP is delivered with a board support package. This includes the integrator configuration itself and device drivers for the major ported operating systems. The device drivers supporting the ARM1136 PXP HdS abstraction are provided for each operating system: Windows CE™, .NET™, Symbian OS™ 7.0 and VxWorks™. Whilst a common driver layer across all operating systems would be desirable, substantive differences between the major kernels offered in the embedded-systems market appear to make this proposal infeasible.

Application software developers can utilize the drivers together with an OS image to run on the ARMulator based Software Development Model (SDM). This provides the basis to port application code and to carry out system performance analysis. The ARMulator model can be extended to provide early functional prototyping of features that will represent the extension and configuration of the platform hardware during a complete SoC integration. The other major use of the ARMulator is to provide an early access model for operating system porting. Since the model is fully functionally consistent with the RTL it greatly accelerates the task of porting a new or derivative OS to the platform.

8.5   Support of system verification and validation

A major issue facing integrators of complex systems IP such as the ARM1136 PXP is that of verifying the final system integration. The ARM1136 PXP product meets this challenge in three ways: (i) by carrying out comprehensive subsystem verification that ensures that the platform itself has been validated under a wide variety of expected corner case scenarios; (ii) by delivering an extendible integration testbench based on the PMAP script outlined in Section 4.3; and (iii) by providing for reuse of the subsystem validation environment at the SoC integration level.

The cornerstone of the system verification and validation support is the XVC standard for verification component extensibility and reuse, as described in Section 7. The XVC methodology specifically supports the first and third verification goals directly, and supports the second by providing test-replay through FRBMs of the PMAP output. Each platform in the PXP family is delivered with a full set of PMAP scripts for the platform. Furthermore, it is provided with a comprehensive set of XVC components that allow the systems integrator to run out-of-the-box test scenarios that ARM engineers used to validate the platform. These verification components are coordinated by an XTSM component that is provided as part of the platform environment.

A conceptual representation of the validation environment provided with a PXP product is shown in Figure 9-12. The capabilities of this environment can be directly used by the systems integrator to run new scenarios, based on the existing supported set of actions of the environment. The PMAP and XVC IP provided with the platform may be re-run by the systems integrator with either a changed or extended address map to verify the integration of any new peripheral.

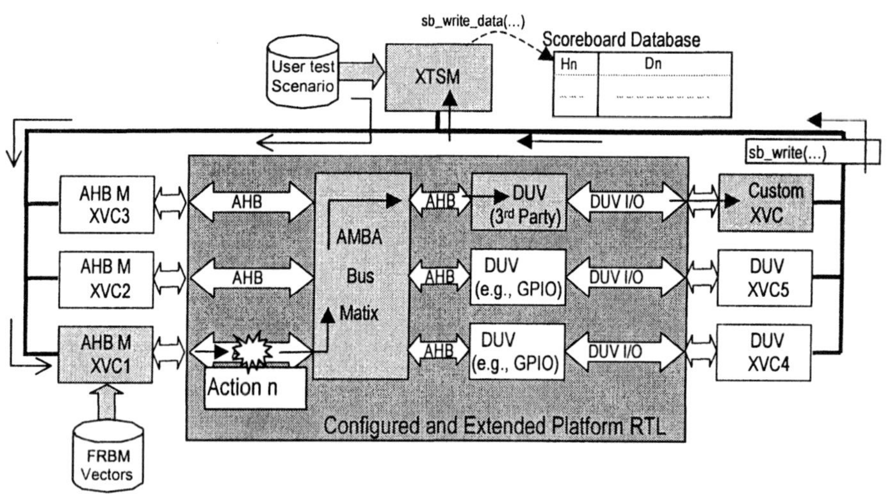

Figure 9-12. Extendible PrimeXsys Platform Verification and Validation Environment

Extension to the XVC verification environment may be done in two ways: firstly through extending the set of stimulus and monitoring conditions within the existing set of platform verification components, and secondly through the addition of new verification components.

In the first case of scenario-extension, consider the Smart Card XVC shipped with the ARM1136 PXP. The basic action set supported by this verification component consists of the ability to respond and set the smart card interface lines. Suppose the integrator wishes to verify a system scenario involving rapid insertions and removal of the smart card (such as might occur when a child 'plays' with the end use system). This can be captured using a new action that toggles the 'card present' line. This action is now available for inclusion in a new scenario. Note that the XVC has been modularly extended, rather than being subjected to unguided modification. This allows for improved support of the verification environment since it is easier to isolate bugs to delivered or extended code.

In the second case of verification environment extension, consider the addition of a new peripheral to the ARM1136 PXP platform. For example this might be the additional of a USB XVC to correspond to the integration of a 3rd-party USB block. By creating a USB XVC that conforms to the standard verification component interfaces of Section 7, this new XVC can integrate with the XTSM and the existing PrimeXsys Platform verification components. This extended verification environment can then be used to create new scenarios that are based on the action capabilities of both the delivered (and possibly extended XVCs) and the new XVCs written specifically for the integration verification environment.

9.   SUMMARY

In this chapter, ARM's lead technical and product marketing minds have provided a broad description of the principles that need to be supported for effective platform based design. These principles have been illustrated with our experience with development and deployment of our PrimeXsys Platform products. As a star-IP provider, ARM is in a unique position to understand the common issues of IP integration that cross the industry. We feel we have come to comprehend the platform-based design problem well, and are providing industry-leading solutions into this space.

This chapter sets the context of platform use models by defining the roles of IP Creator, IP Licensee and IP User. We have defined the set of platform-layers and associated use-models for application software development, middleware and systems micro-architecture development, and hardware integration. We have expressed how both the IP exchange in support of a platform and internal design-refinement flows are enabled by: modeling and prototyping views of the platform, standard interface development and compliance, and cross-hierarchy verification and validation support.

ACKNOWLEDGEMENTS

The authors would like to thank the many engineers, managers and product support personnel involved in the ARM PrimeXsys Platforms development. In particular, we thank Jonathan Morris and Pete Aldworth for their review of the chapter, and their critical input into defining the ARM platform products.

REFERENCES

1. ARM ePublication, "PrimeXsys™ Platforms Extendable Platform Architecture", http://www.arm.com/armtech/PrimeXsys/.
2. VSIA homepage. http://www.vsi.org/.
3. Thorsten Grötker, Stan Liao, Grant Martin and Stuart Swan, *System Design with SystemC*, Kluwer Academic Publishers Group, 2002. ISBN: 1-4020-7072-1.
4. Open SystemC Initiative homepage http://www.systemc.org/.
5. Accelera homepage. http://www.accelera.org/.
6. ARM ePublication, "AMBA™ The defacto standard for on-chip bus", http://www.arm.com/armtech/AMBA/.
7. Symbian OS application Software Development Kits, http://www.symbian.com/developer/downloads/.
8. Wander O. Cesário, et al. "Multiprocessor SoC Platforms: A Component-Based Approach", *IEEE Design & Test of Computers*, pp. 52-63, Vol. 19, No. 6, 2002.
9. Wilson, R. "SoC bus war fizzles", *EE Design*, July 1 2002. http://www.eedesign.com/.
10. A. Haverinen, M. Leclercq, N. Weyrich, and D. Wingard, "SystemC™ based SoC Communication Modeling for the OCP™ Protocol", OSCI Technical Paper, October 2002. http://www.systemc.org/.
11. C. Lennard, D. Lin, K. Ho, and V. Chandran, "Collaborative Development of Bluetooth IP: Fundamentals of System-Modelling in Complex IP Creation and Reuse", *Proceedings of IP 2000 Bluetooth Conference*, pp. 329-344, October 2000.
12. B. Bailey, G. De Jong, P. Schaumont, and C. Lennard, "Describing Interfaces Using the VSI System Level Interface Behavioral Documentation Standard", *Proceedings of the Forum on Design Languages*, pp. 221-230, September 2000.
13. A. Nightingale, "Testing for AMBA TM Compliance", *Proceedings of the 14th. IEEE ASIC/SOC Conference*, September 2001.
14. J. Connell and B. Johnson, "Early hardware/software integration using SystemC 2.0", *Embedded Systems West Proceedings*, Class 522, 2002.
15. Jerzy Rozenblit and Klaus Buchenrieder, *Codesign: Computer-Aided Software/Hardware Engineering*, IEEE Press, 1995. ISBN: 0-7803-1049-7.
16. Filip Thoen, "Prototyping Technologies for Early Embedded Software Development", *Information Quarterly*, pp. 17-21, Vol. 1, No. 1, 2002.

17. A. Bruce and J. Goodenough, "Reuseable Hardware/Software Co-verification of IP Blocks", *Proceedings of the 14th. IEEE ASIC/SOC Conference*, September 2001.
18. J. Goodenough, A. Bruce, A. Nightingale, P. Bates and G. Budd, "A Unified Validation Methodology for System Level Co-Design and Co-Implementation", *Proceedings of the 14th. IEEE ASIC/SOC Conference*, September 2001.
19. ARM ePublication, "ARM11 Family: ARM1136J-S and ARM1136JF-S", http://www.arm.com/armtech/ARM11/.

Chapter 10

REAL-TIME SYSTEM-ON-A-CHIP EMULATION
Emulation Driven System Design with Direct Mapped Virtual Components

Kimmo Kuusilinna[1,2], Chen Chang[1], Hans-Martin Bluethgen[3], W. Rhett Davis[4], Brian Richards[1], Borivoje Nikolić[1] and Robert W. Brodersen[1]
[1]University of California at Berkeley, Berkeley Wireless Research Center; [2]Tampere University of Technology, Finland; [3]Infineon Technologies AG; [4]North Carolina State University

Abstract:
The productivity gap between the designer and the opportunities for silicon integration places increasing pressure on system verification in particular. A comprehensive design flow for digital systems from high-level algorithmic specifications to FPGA-based emulation and final ASIC implementation is presented. The design is entered only using a component library with predictable performance, therefore, enabling rapid system development and easing the verification burden. Hardware emulation, from this description, enables rapid prototyping of large systems where gate-level simulations are impractical. The primary goal of the emulator is to support design space exploration of real-time algorithms. The design environment is customized towards low-power and data-flow dominant architectures, particularly focusing on applications related to wireless communications. The design process of a 1 Mbit/s transmission system is explored, demonstrating the design convenience and the early performance analysis.

Key words:
Electronic design automation, hardware emulation, rapid prototyping, field-programmable gate-array.

1. INTRODUCTION

In order to most efficiently implement system-on-a-chip (SoC) designs, it is desirable to employ the expertise of the algorithm developers throughout the design process and across the levels of design hierarchy. Designers tend to choose a form of design entry that they are most familiar with, which often results in the desire to use a standard software programming language. It is also of ultimate importance to provide a feedback path to the algorithm developer about the feasibility and performance of the implementation. Finally, it is also often desirable to evaluate the performance of the target system through a real-time prototype.

Satisfying these constraints using conventional approaches leads the SoC developer to a software solution with complexity estimates made by monitoring the execution run time. The design can then be prototyped on a processor that employs the same instruction set. However, the integration of more optimized hardware accelerators becomes very difficult because of the performance mismatch between a software implementation and hardware acceleration.

Programmable processor-based solutions are far less energy efficient than direct mapped algorithms on a silicon chip [4]. In many cases, the processor performance does not meet the demand for real-time prototyping or emulation. Examples of this include advanced modulation and coding schemes, such as implementation of space-time codes or iterative decoding algorithms. In such cases, it is difficult to achieve the optimal performance of the algorithm on the emulator. Another example is the verification of a MAC (Medium Access Control) layer implementation for wireless LANs.

This Chapter describes a different approach which satisfies the above requirements of flexibility in the system specification stage, but is able to drive an implementation using highly optimized architectures which are energy and area efficient, and execute orders of magnitude faster than is achievable with a processor-based solution.

The basic approach is to take a fully parallel description of an algorithm using a timed data-flow description and then to directly map it into hardware using libraries that provide resource requirements. The libraries can then be mapped to two targets: an ASIC (Application Specific Integrated Circuit) implementation for the actual SoC integration and an FPGA (Field-Programmable Gate-Array) realization for real-time prototyping. In the particular implementation presented here, the timed data-flow description used is Simulink from Mathworks [10], which is a common environment for communication system developers. By providing accurate resource estimates at this high level, the system designer can perform architectural optimization with accurate knowledge of the final implementation costs. The resulting system is then automatically mapped to a real-time prototype on a large FPGA array shown in Figure 10-1. This prototype is bit-equivalent and cycle accurate to an integrated SoC implementation that can then be generated from the same input description.

The direct mapping of the high-level block diagram description has the advantage that automatic transitions between different CAD descriptions of the design are possible due to the straightforward conversions. Design verification is easier because difficulties in lower-level descriptions are conveniently associated with elements of the top-level design. Furthermore, this means that most, if not all, of the design decisions can be raised to the top-level. With direct mapped designs, the algorithm details are explicit, which forces the algorithm designer to be involved in the hardware design on a high level. The designer can quickly see the impact of architectural decisions on the actual silicon implementation.

This Chapter is organized as follows. Section 2 describes the fundamental approach taken for the SoC design. Section 3 discusses the basics of system design and the Berkeley Emulation Engine (BEE) integration into the system design flow. In addition, virtual components and the library development methodology are examined. In Section 4, the implementation paths to both FPGA-based emulation and ASICs are explained, focusing on the BEE architecture and some of its inherent capabilities to support contemporary logic design. Section 5 is an application example of a system building block: a 1 Mbit/s transceiver. Finally, Section 6 concludes the Chapter.

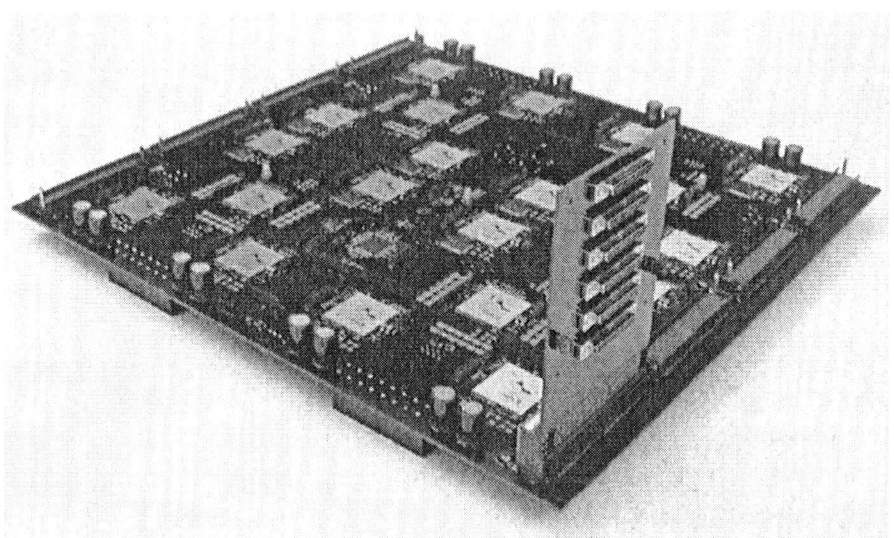

Figure 10-1. The BEE main processing board with one riser card.

2.   DESCRIBING SOC DESIGNS

A "single-chip radio" SoC requires analog circuitry with RF analog and baseband circuitry, A/D and D/A converters, memories, and hard-wired digital signal processing, each implemented with the highest possible performance at the lowest possible power. In this Chapter, we will concentrate on the digital part of the problem and, in particular, a strategy that allows real-time emulation of highly optimized architectures.

Many radio chip design flows can be broken into four stages: specification, architecture definition, front-end, and back-end design. As the first stage, specification generally includes a description of the algorithms and protocols for the digital portion, and requirements for the analog portion such as noise figure, phase noise, and distortion. The second stage is typically where the IC designer explores the design of the chip architecture. An architecture is a general plan of the functional organization and the physical configuration of the design. For SoC design, this includes a preliminary plan of subsystems and blocks as well as the plan of the global signals that will be processed, stored, and carried from block to block. A particular instance of an architecture, that is a micro-architecture, should refine these plans for spatial locality. The further a signal must be carried on the chip, the more likely it is to incur delay or be corrupted by noise and coupling effects. The third stage, front-end design, includes the complete specification of all logic functions on the chip, generally as RTL logic, and models for all analog blocks. Finally, back-end design includes the mapping of the front-end design to transistors and making the final mask patterns to be used in fabrication.

Chip projects generally begin with a specification and proceed sequentially through the three later stages of the design flow, frequently with a different team working on each aspect. The success of this approach is that it allows the design flow to be broken up into many parts, which can proceed more or less independently. Changes to this kind of a flow, and therefore for each new generation of wireless systems, are usually relatively small so that the balance between the stages is not disturbed. Micro-architectures change slowly to retain predictable performance, making multi-level design optimization difficult. However, it is preferred for the algorithms to drive the implementation decisions instead of letting implementation dictate algorithms.

An important aspect of a design flow is its ability to leverage technology optimization at each abstraction level, enabling global design optimization. For example, at the micro-architecture level, altering the number and type of operations and adding parallelism should be possible. At the architectural level, one should be able to simplify interfaces between blocks, adjust supply voltage, add buffers and caches to increase throughput and to remove buffering constructs to reduce power. At the front-end level, switching off unused circuitry, resource sharing to reduce area, elimination of resource sharing to reduce power, and pipeline re-timing to minimize cycle time are important. For the back-end, careful floor-planning to reduce interconnects in critical paths, reducing noise to sensitive circuitry, and resizing transistors to reduce power or increase speed are issues to consider.

Our approach has been to choose a single environment, which acts as a common language for analog, digital, and algorithm designers. The common language and simulation paradigm allow the effects of design decisions to be quickly checked against the rest of the design. Simulink was chosen as the common environment because it is a good compromise among the different design environment requirements, namely it has the ability to describe designs with both analog and digital implementation targets and support for data-flow and state machine (through Stateflow) design entry. Since Simulink has a structural rather than a procedural description, the system descriptions inherently represent the basic parallelism, which can be exploited by concurrent hardware. Both analog and digital subsystems can be described in Simulink. Analog circuits are typically modelled with abstractions for non-idealities, such as baseband RF front-ends that include modeling of phase noise in the VCO (Voltage Controlled Oscillator), circuit noise in the LNA (Low-Noise Amplifier), and distortion in the mixer. For digital designs, datapaths are described using a fixed-point block set, and control logic is described using the Stateflow finite state machine package. Typically, these models use discrete time instead of continuous time. Therefore, Simulink allows the description of a class of synchronous, mixed-signal, heterogeneous systems.

The micro-architecture is defined by mapping the Simulink blocks to hardware. The blocks correspond to hard or soft macros. Stateflow blocks and simple look-up tables correspond to RTL code, which will be synthesized to a standard-cell netlist. Other blocks correspond to semi-custom module generators that create parameterized blocks for datapath circuits such as adders and multipliers. Thus, to see the effect of an optimization is to see how these well-understood blocks interact.

The real challenge for this design flow is to provide automation that is seamless enough to let all designers understand the interactions of the blocks, regardless of their CAD expertise. The goal is to make the entire flow simple enough for one person to take a design from specification to mask layout, and quickly receive the feedback about the physical implementation.

3.   SYSTEM-LEVEL DESIGN FLOW

The single most critical aspect of this design flow, besides the functional correctness, is the speed at which the designer can try out different implementations. Early estimates of performance allow rapid design evolution, avoiding time-consuming design steps such as synthesis of designs that are likely to be rejected. In addition, when moving between different design abstraction levels, the designer should not need to manually verify the correctness of the design. For example, manually entering test benches is quite cumbersome, and should be automated.

The direct mapped design strategy using virtual components benefits from architectures with only a limited amount of dynamic control. One model of computation for such designs is the synchronous data-flow model, which is the primary way we choose to interpret the top-level design descriptions. This model is convenient for many digital signal processing applications since it captures the design in a concise but relatively unambiguous manner. Therefore, the path from high-level description to hardware is not broken.

Two issues merit special emphasis when designing with the synchronous data-flow. First, deadlocks or unresolved signals cause simulation problems, and may result in an ambiguous hardware implementation. These situations can arise, for example, from poorly defined initial conditions and feedback loops. Particularly, feedback loops without explicit delays can cause problems. The solutions are explicit initial conditions and breaking the feedback loops with delay elements. Second, in designs with multiple clock speeds, the data rates must be kept consistent at the clock domain boundaries. The input gateways in the design keep track of the original data rates and this information is then propagated through the whole design. When crossing to or from different clock domains, the data must be converted by a corresponding up-sample or down-sample component.

3.1   Fundamental Concepts in the Design Flow

A general system-level development flow is depicted in Figure 10-2. Beginning from the desired application functionality and the specifications along with the available architectural resources, the design is refined into an implementation. The role of early performance analysis is again emphasized at all abstraction levels as the key to functionally correct and high performance implementations. A finished design contains all the representations from all the available abstraction levels.

The block arrows in Figure 10-2 represent the primary information flow. The black arrow is a design flow internal feedback path, which is used to accept or reject particular implementations. The gray arrows signify relationships that are more abstract. The results from performance analyses and the realized designs influence the designers to innovate new applications and specifications. In addition, designs can become the basis for micro-architectural structures and new library components.

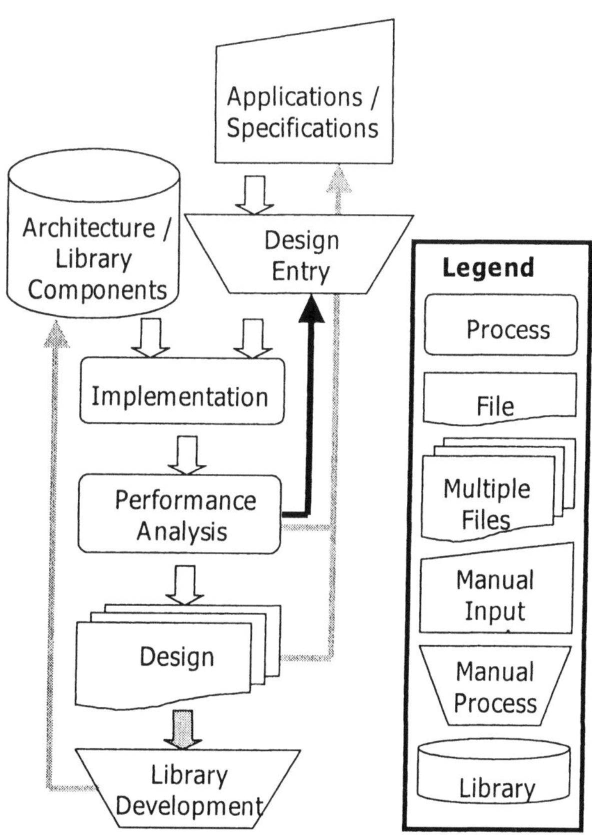

Figure 10-2. System design flow [7].

For this system design flow to work, it is essential that the designer is not too constrained by the available architectures and components. If the architectural platforms and components in library are fixed, either the design space or the productivity is severely limited, depending on the size of the components. Therefore, parameterizable virtual components are a natural compromise. As with most other flows, this approach is viable only if sufficient libraries exist and they are easily extensible.

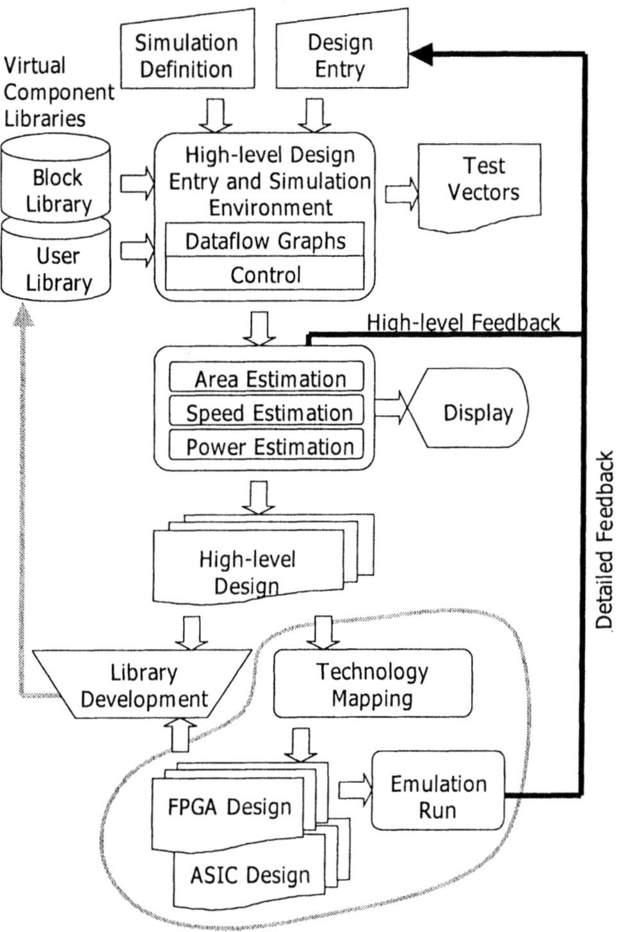

Figure 10-3. Emulation driven system design flow.

3.2 System Design Utilizing BEE Emulation

The speed and the quality of information in the internal evaluation loop, depicted in Figure 10-2, is one of the primary problems in system design. Figure 10-3 depicts the BEE system design flow where the performance analysis is predominantly either done at a very high level or is based on hardware emulation. Both methods are very fast compared to detailed simulations and the emulation results provide additional confidence form the hardware level for the design decisions.

For BEE-based designs, the block library is mainly based on Xilinx components like registers, logic gates, adders, and multipliers. Higher-level IP libraries are used for components like FFT and FIR. At the high-level, design capture and simulation environment is based on Mathworks Simulink (version 6.5) simulator [10] combined with Xilinx System Generator (version 2.2) [11].

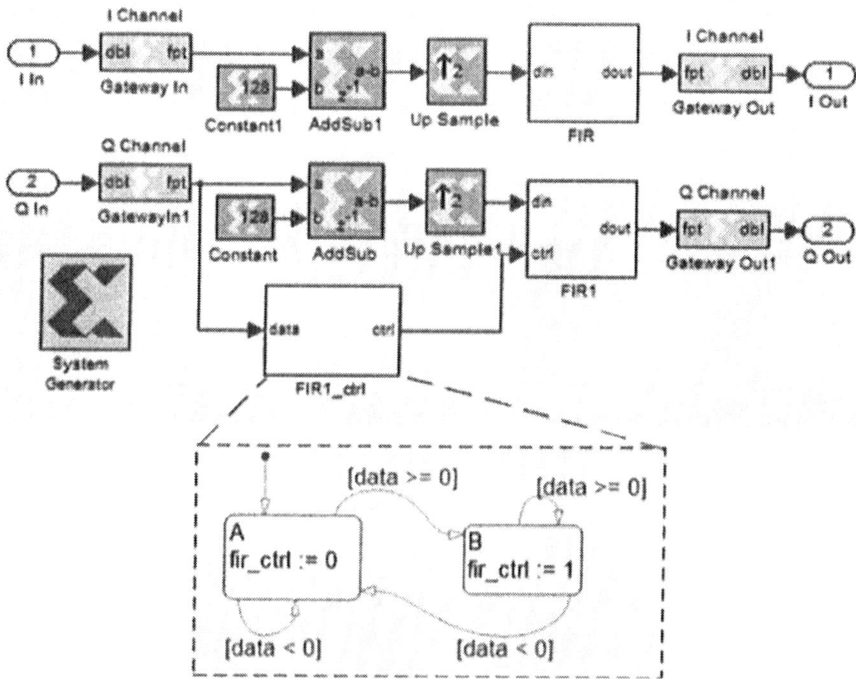

Figure 10-4. Design entry with both datapath and control elements.

This collection of tools and libraries forms the basic framework for our top-level infrastructure. Our development effort is, therefore, limited to adding missing functionality, like high-level estimation, and extending the component library. The environment also provides automatic VHDL test bench generation. The estimation step is built on top of these tools and, in the case of ASIC designs, on additional technology characterization. Figure 10-4 depicts an example of a typical design entry, which contains direct mapped components like *AddSub* and *Up Sample*, hierarchical subsystems like *FIR* and *FIR1*, and synthesizable control like *FIR1_ctrl*. The control is described as a state machine in Mathworks' Stateflow environment.

Exposing the designer only to the high-level design environment is an important concept in this flow. That is, all the design decisions should be made at the top-level. In addition, feedback from the underlying flow steps needs to be presented in a form that the designer can easily process and convert into alternate design decisions and optimization goals.

User decisions are divided into functional, signal, circuit, and floorplan categories. Routing problems are typically solved automatically. Functional decisions relate to the system responses to external stimuli and system behavior in general. Signal decisions deal with physical signal properties particularly with word lengths. Signal timing is specified as the minimum clock frequency of the circuit. Circuit decisions specify the transistors to implement each subsystem and the overall subsystem architecture. A floorplan indicates pad placement and optionally the physical locations of functional units. In addition, high-level floorplanning on subsystem level is utilized for manually partitioning the design for the FPGAs on a BEE Processing Unit (BPU).

Area, speed and power are the three traditional focus points when considering the quality and performance of a design. Providing high-level feedback to the user based on these criteria allows the designer to focus on the architectural decisions. The design area is expressed as a scalar number that is proportional to resource utilization, based on the component types and parameters. In case of FPGAs, the number is related to *slices*. A slice is the Xilinx term for a collection of two four-input look-up tables, two registers, and carry & control logic. In case of ASICs, the number is based on Synopsys Design Compiler area estimates for the target technology. In addition to top-level evaluation, the user can request area estimation for any of the subsystems. The operating speed is estimated from a table-lookup of the maximum delay of each library block, previously obtained from the FPGA timing analysis tools, or measured from the emulated system. The power estimates are based only on the number and the types of the components and the target clock frequency.

This feedback should primarily translate into functional and signal decisions and optimizations since the system architecture is the part of the design where decisions based on accurate information count the most and the greatest savings in power and area are available. The number of circuit decisions should be minimal since these issues are dealt with in the library development. Floorplanning inside the chip is typically not exposed to the user, but partitioning the subsystems to each of the FPGAs is the responsibility of the designer and potentially affects the global system architecture.

Design for emulation has several high-level, global optimization goals. The number of FPGAs in the design should be minimized since this typically speeds up the design and tends to limit the number of signals between the FPGAs. The external interfaces either to the emulator itself or between the components inside the emulator are potential performance bottlenecks for any emulator. Virtual wires [1] can be used to overcome the physical wire number limitations, but these methods incur a speed penalty. Good design practices like pipelining, registered outputs, and restricting the logic depth should be followed, particularly for off-chip signals. [3]

4. SOC IMPLEMENTATION PATHS – EMULATION AND ASICS

4.1 Berkeley Emulation Engine

Hardware emulation is usually seen as a form of *rapid prototyping* where a physical system that implements an algorithm is developed using an existing HW framework to speed up the design process. In addition, the term hardware emulation typically refers to synthesis-based design methods and to a fabric of configurable logic. Preferably, the emulation runs at the same speed as the final product.

The Berkeley Emulation Engine was built to allow rapid algorithm exploration in the hardware level. In addition, many designs that are difficult to simulate due to their size and complexity can be designed utilizing BEE. A class of real-time applications from the wireless communication domain [3,5] can be emulated, providing real-time feedback on the effects of algorithm optimizations. For example, the results of a quantization selection can be observed. After the definition of the algorithm, the description can be utilized in both the emulation environment and in the final ASIC implementation, while maintaining cycle-accurate and bit-true correspondence.

4.1.1 Direct Mapped Designs and the BEE Architecture

Direct mapping implies that the top-level design elements already explicitly specify the micro-architecture with the cycle-accurate behavior. For example, an 8x8-bit multiplier is instantiated as specified in Simulink, with the pipelining depth as specified in the parameters, all the way to the silicon implementation. This facilitates the development of fast CAD tools and the FPGA and ASIC implementations are functionally the same. Direct mapping is especially suited for designs with a high-level of parallelism or optimization goals that emphasize low-power with stringent performance specifications.

The underlying goals of the Berkeley Emulation Engine hardware development were to provide a large, unified, real-time emulation platform for data-flow-centric designs. Combining these requirements with the direct mapped design approach resulted in a *Two-Layer Mesh* routing architecture for BEE. Balancing between fast local interconnections that cross a FPGA chip boundary and the global connectivity is perhaps the most crucial HW emulator design parameter. BEE is optimized towards local connectivity. In other words, the number of *hops*, chip-to-chip connections, to neighboring FPGAs is minimized. Figure 10-5 depicts the BEE routing architecture.

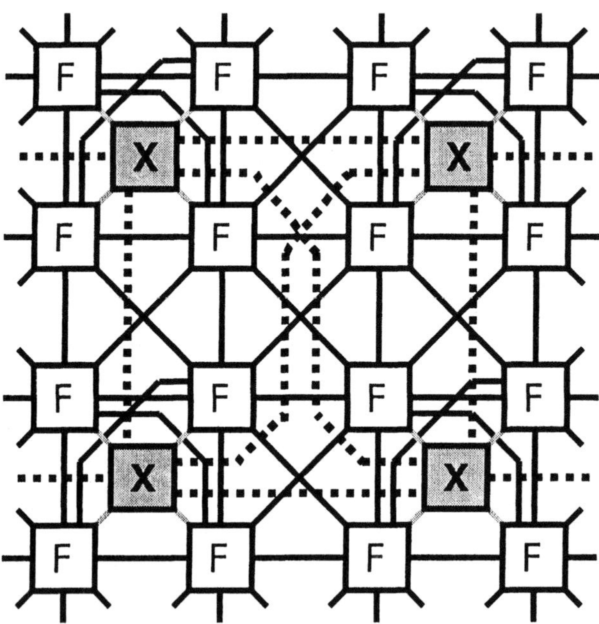

Figure 10-5. BEE routing architecture. The solid lines are the first-level mesh and the dashed lines the second. Each of the squares represents a FPGA with those marked with 'F' connect to the first-level mesh and the ones marked 'X' connect to the second level.

An aggregate of FPGA chips on a printed circuit board, in addition to all the supporting equipment, is called a *BEE Processing Unit* (BPU) and it is large enough to emulate systems with up to 10 million ASIC equivalent gates. Physically, this is implemented as a *Main Processing Board* (MPB) with 26 signal and power layers, 20 Xilinx VirtexE 2000 chips, and 16 SRAMs, each with a capacity of 1 MB for data buffering purposes. Larger emulation systems can be constructed by connecting multiple BPUs together using the external I/Os. Designs implemented on the BEE platform cannot be very power efficient due to the inherent limitations of the FPGA technology. However, retargeted to an ASIC technology the same design could achieve a very low-power behavior.

Both the design flow and the BEE hardware support two low-power design techniques, multiple clock domains within a single design and clock gating. Within each FPGA, the primary clock frequency can be multiplied by 2 or 4 and divided by 1.5, 2, 2.5, 3, 4, 5, 8, or 16 using a delay-locked loop. Other frequencies are available if logic is used to divide the clock. In addition, all the FPGAs have three additional clock inputs, which can be used to input clocks that are completely independent from the primary clock. The FPGAs do not support real clock gating, but the registers have a clock enable input that can be utilized to emulate the behavior.

4.1.2 BEE Emulation

Figure 10-6 depicts the BEE hardware infrastructure and the information flow within the emulation system. A BPU has a separate connection to the Host Server via dedicated Ethernet. The server is responsible for configuring the system and for allowing it to be remotely accessed from client workstations. Analog radio front-ends can be used to form complete transmission systems.

An integrated Single Board Computer (SBC) enables a BPU to be connected to Ethernet. The information flow between the Host Server and the BEE system can begin after the user has generated the necessary design files utilizing the BEE design flow. The design files are sent to each BPU through the Ethernet and stored in memory or the hard disk of the SBC. Finally, the user issues commands to the SBC, instructing it to either configure the BPU or read back information.

The SBC connects to the 20 FPGAs on an MPB through a configuration FPGA, which mainly serves as a bi-directional signal multiplexer between the 16 general-purpose I/O lines from the SBC to over 100 control signals on a MPB. In addition, the off-board main power supply system is controllable through this link. All control functions on a BPU can be controlled from the SBC. The functions are divided into the following categories: programming of the processing FPGAs, data read-back from the processing FPGAs, clock domain control, power management, and thermal management.

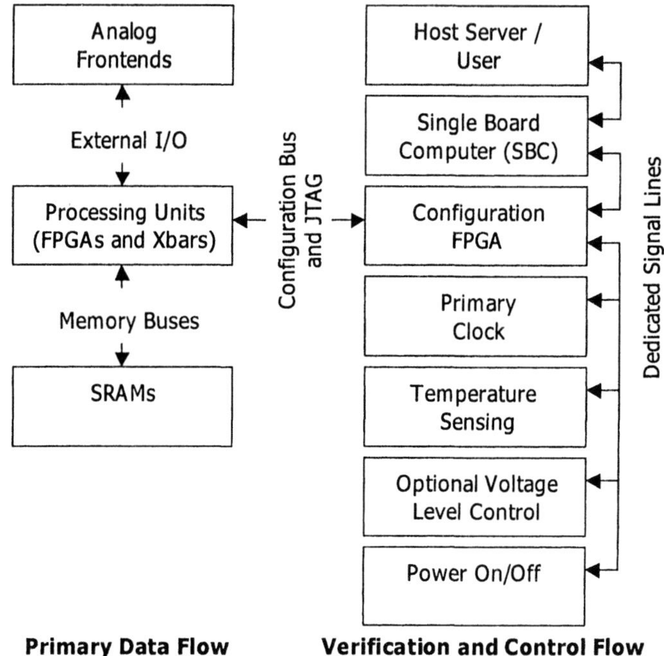

Figure 10-6. BEE information flow and emulation setup.

On the MPB, less than 2% of the total signals on the board are directly accessible through probing headers. Using FPGA programming, internal signals could be routed to these headers for direct probing with a logic analyzer. However, this is not enough for practical hardware debugging. Therefore, a software-based digital logic analyzer solution is used as the primary debugging tools for the MPBs.

Xilinx ChipScope Integrated Logic Analyzer (ILA) [11] cores can be inserted into the design where they act as tiny logic analyzers at run-time. The ILA records the values of the monitored signals and transmits these through the JTAG interface back to the host workstation. The ChipScope software collects data from different ILA cores, which can reside on different FPGAs, and combines them onto a signal waveform display. In addition to the ILA cores, which use the on-chip BlockRAM as data storage, the external SRAM could be used for synchronous signal recording.

4.1.3   Prototyping Concepts for BEE Hardware Emulation

The BEE emulation is based on the concept of functional and cycle-level equivalence between the emulation and the final ASIC implementation. This means that signals have identical behavior in both implementations, but the underlying low-level hardware architecture may be different. And as noted before, there is only one top-level description and simulation of the system.

The design flow automation and the emulation facilitate rapid prototyping where the design space of requirements, alternate specifications, and implementation feasibility are explored before committing to the final design and its optimizations.

In addition, this *concept-oriented prototyping* [9] is well suited for hardware-accelerated simulation and computation. Computation acceleration is particularly attractive for computationally intensive but parallelizable algorithms. The rapid hardware acceleration path can be utilized to build specialized hardware systems to solve problems that would otherwise have to be addressed with slow computer programs because of costs related to time, manpower, and complexity of building dedicated hardware. Simulations can be accelerated either with synthesizable test benches or by offloading some of the simulation computations to the emulator. The latter option is particularly efficient since usually there is no need to emulate the simulation computations, but the simulated hardware can be simply run on the emulator.

The high-level simulations and rapid prototyping verify the functionality and physical behavior of the design. The goal is to gradually build confidence on the high-level simulations and prototype large systems on the BEE emulator. Particularly designs that are too large and complex to be conveniently simulated leverage the resources available in BEE. However, special circumstances, like the library development, may require even finer grain design verification.

BEE excels in real-time *in-circuit verification*. That is, physical components from the target design, typically radio front-ends, are added to the emulation system. Real-time emulation runs increase the confidence that all the sub-designs are integrated correctly and systems can be tested in their real operating environment. Furthermore, in many signal processing applications, the processing speed affects the perceived quality of the design, thus, increasing the value of real-time experiments.

4.1.4 Designing for Hardware Emulation

Many hardware emulators follow a generic design flow beginning from behavioral system design and behavioral synthesis to emulator specific operations like emulator partitioning, logic synthesis, and technology mapping [3,8]. The end result is the emulation run. Due to the direct mapped nature of a typical BEE design, behavioral synthesis is unnecessary and the need for complicated logic synthesis is minimal. However, a lot of emulation technology dependent design flow steps are analogous for general emulators and BEE. The BEE technology mapping part of the BEE flow is depicted in Figure 10-7.

Partitioning for heterogeneous resources like the BEE emulator is known to be a hard problem. Approaching this problem from the FPGA point of view has been documented, for example in [6]. The system-level routing architecture has a profound influence on this design phase and the typical designer has a lot of *a priori* information on the layout of the design. Therefore, the high-level partitioning is left for the user. Routing, as mentioned, is automatic on all levels of the design. The partition information also directly indicates which FPGAs should receive which bitstream. This information is called the Emulator Configuration.

The Xilinx System Generator and the Integrated Synthesis Environment (ISE) automatically execute the internal phases of technology mapping for individual FPGAs. The design netlist is inferred from the high-level design. Core components are generated based on the virtual component libraries; thus, a core is an instance of a virtual component that has a fully specified implementation. Synthesis is applied for the parts of the design requiring it and VHDL test benches are generated. The back-end of the flow partitions, maps, and routes the FPGA.

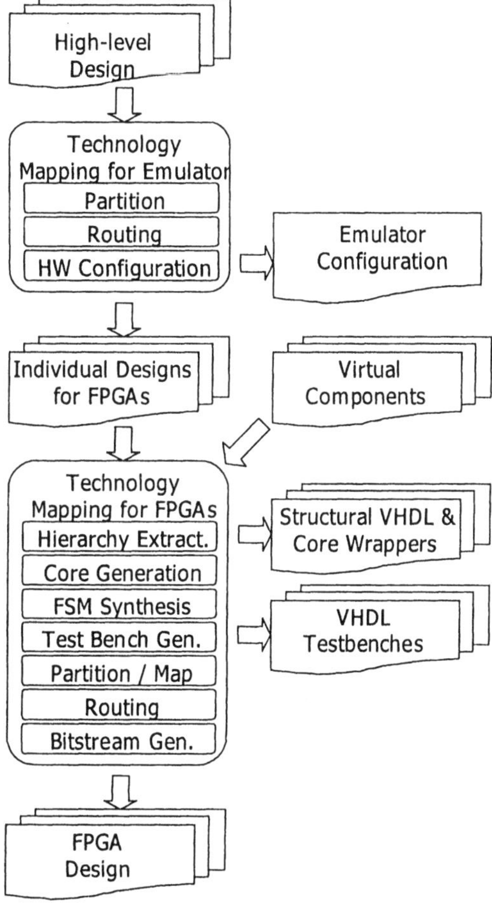

Figure 10-7. HW emulator technology mapping.

4.2 Designing for ASICs

An ASIC implementation is possible after the design has been evaluated and approved using BEE hardware emulation. Our ASIC flow is based on the work done in a previous project called Simulink-to-Silicon Hierarchical Automated Flow Tool [4]. The virtual components for ASICs are in the form of parameterizable Synopsys Module Compiler descriptions. The ASIC technology flow is depicted in Figure 10-8.

The Frontend Technology Mapping utilizes the System Generator generated design hierarchy, VHDL core wrappers, and test benches. The design is imported to the Synopsys synthesis framework where cores are instantiated and synthesis is done to the appropriate parts of the design. In addition, a boundary scan chain can be added to the design at this stage. The Synopsys framework outputs a hierarchical netlist.

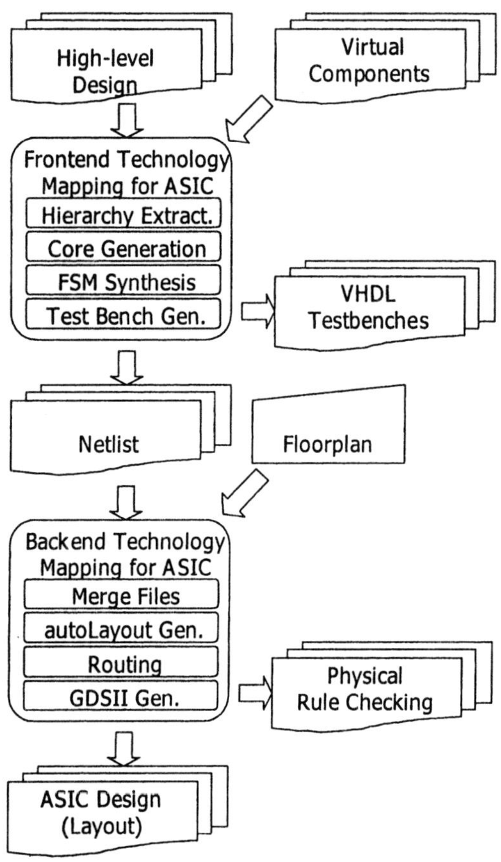

Figure 10-8. ASIC technology mapping.

The Backend Technology Mapping is based on the Cadence tool suite. A floorplan can be entered manually if desired. These inputs are merged and the layout is optimized. The final major design refinement phase is the automatic routing. In addition, design rule checks are run against Spice netlists and GDSII.

5. CASE STUDY: A 1 MBIT/S NARROW-BAND TRANSMISSION SYSTEM

Digital communication circuits are an application domain which is particularly suited for our design environment, as was discussed earlier. A block diagram of a typical digital communication system consisting of a transmitter, a channel, and a receiver is depicted in Figure 10-9. To demonstrate the features of the design environment, a simple 1 Mbit/s narrow-band transmission system has been designed.

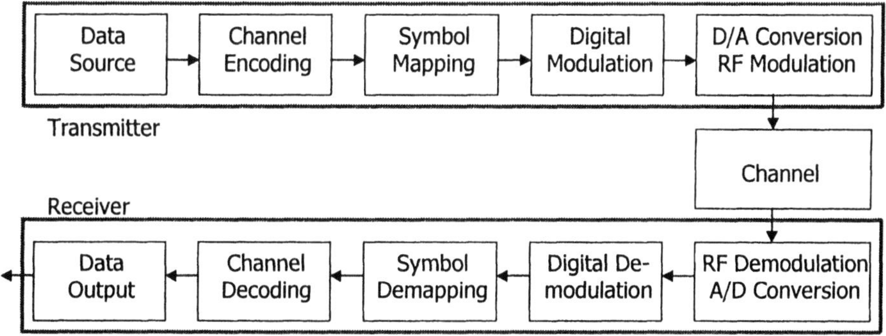

Figure 10-9. Block diagram of a basic digital communication system.

The design is emulated with built-in test vector generation and in an in-circuit mode with analog radio front-ends (2.4 GHz transceiver). Following this verification, an ASIC layout is generated using the same description. In the following, the basic design steps are described for the transmitter portion of the transmission system. A table, collecting the emulation results summarizes also the channel model and the receiver part of the design.

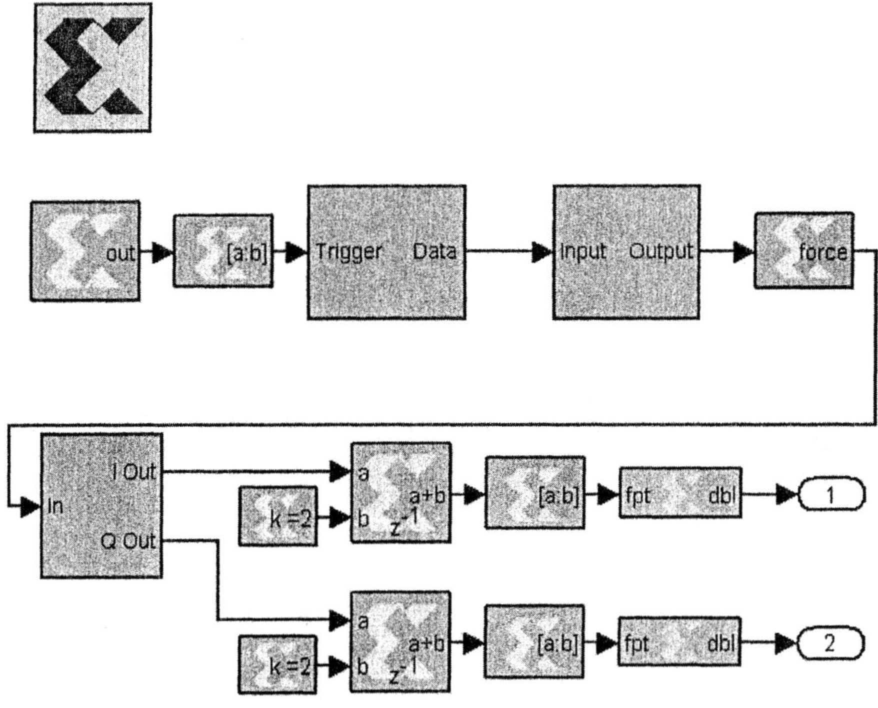

Figure 10-10. The top-level block diagram of the 1 Mbit/s transmitter.

5.1   DQPSK 1 Mbit/s Transmitter

Figure 10-10 depicts the top-level Simulink / Xilinx System Generator design entry for the transmitter. The *Data Source* block generates a data stream and performs the mapping to symbols. After up-sampling and low-pass filtering, the signal is modulated onto a low intermediate frequency. The solid gray blocks contain subsystems and the rest, like *AddSub1*, are System Generator blocks that have a direct parametrizable hardware implementation.

The transmission system operates frame-based and utilizes a differentially encoded pseudo-noise sequence for frame detection. The transmitter as well as the entire transmission system have been designed for a master clock frequency of 32 MHz.

5.2   High-Level Analysis and the Emulation Run

For this experiment, only one BEE Processing Unit is required. The emulation run is arranged so that the transmitter resides on one FPGA, the transmission channel is modeled with another FPGA, and a third FPGA functions as a receiver. Table 1 tabulates general statistics for the whole system. The results show that practical systems can be constructed with this blockset in reasonable time. A significant portion of the design time was spent by the designer familiarizing himself with the blockset and the design flow in general.

Table 10-1. Some general properties of the emulated system.

	No. of design objects	Types of objects	Max. levels of hierarchy	Design time
Transmitter	133	22	5	1.5 weeks
Channel model	15	8	1	1 day
Receiver	209	26	4	3.5 weeks

Table 2 tabulates the high-level (Simulink) area estimates, the final resource usage, runtimes for the estimation and synthesis, run times for the software, and the estimated maximum clock speed after synthesis, placement, and route. The results show that the high-level area estimation is relatively accurate and the run-time to achieve these results reasonable.

Table 10-2. Transmission system implementation data.

	Transmitter	Channel	Receiver
Est. slices	902.5	3393	2373
Final slices	933	2347	2802
Est. LUTs	1433	4144	3525
Final LUTs	1073	2366	3819
Est. flip-flops	1521	6784	4407
Final flip-flops	1430	4257	4305
Est. block RAMs	3	0	9
Final block RAMs	3	0	9
Runtime for est.	27 s	33 s	81 s
Max. clk frequency	57 MHz	62 MHz	33 MHz
Synthesis runtime	2:42 min	6:54 min	9:02 min

In this design, the average error in the number of slices is 17% and the average error is 25% for look-up tables (LUTs) and 15% for registers. The BlockRAMs were always reported accurately. The estimation error is considerably larger for the channel implementation due to a small number of components, and problems in estimating the size of the FIR filter in this design. The average time for estimation is 13% of the time required for running the synthesis and the estimation should even do comparably better with larger designs.

The FPGA utilizations in this design range between 5 and 15%, which were intentionally kept low for future design expansions with the opportunity to maintaining the system-level partitions. In addition, partitioning early for multiple chips facilitates the division of work between several designers. In general, the target utilization should not exceed 80%, which alleviates routing congestion and helps to achieve the real-time requirements. In this case, the emulation was able to run at 32 MHz, as required by the design. Figure 10-11 depicts a portion of the emulation results namely the received data and some of the receiver control signals.

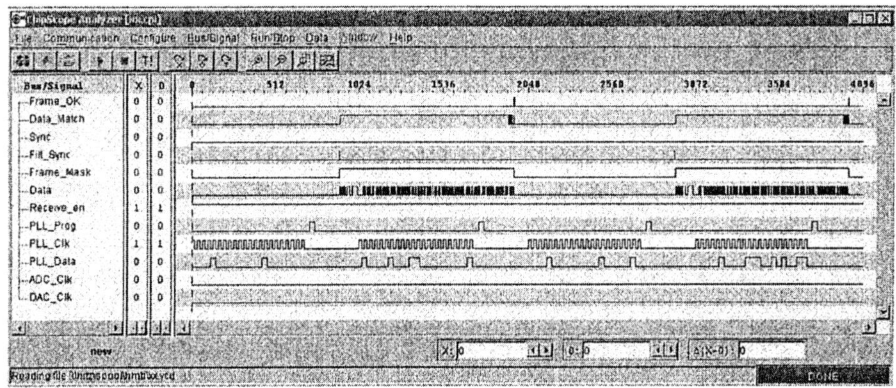

Figure 10-11. Signal waveforms from the emulation.

5.3　ASIC Implementation

Running the transmitter through the ASIC flow took 56 minutes of processor time on a 400 MHz Sun UltraSPARCII and the resulting layout is depicted in Figure 10-12. The core area is 0.28 mm^2 with a utilization factor of 0.34, thus being heavily pad limited. The estimated maximum clock speed is 100 MHz, which satisfies the 32 MHz target speed. The dynamic power is estimated to be 0.611 mW and leakage power 0.016 mW. The target technology is the ST Microelectronics 0.13 μm CMOS process with low-leakage standard cells.

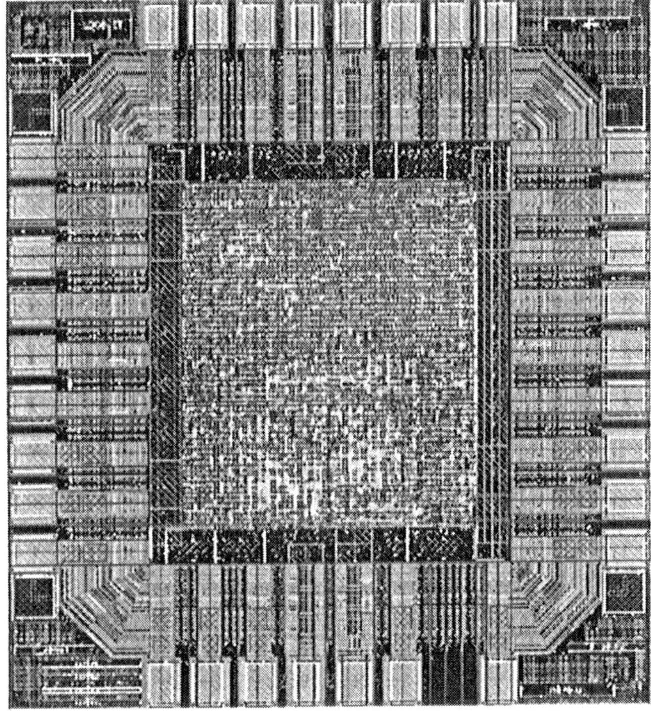

Figure 10-12. Layout of the transmitter.

6.   CONCLUSIONS

A methodology for the development and rapid prototyping of data-flow dominant hardware system designs was introduced. The design flow is based on direct mapped virtual components, which simultaneously allow a high degree of designer productivity and predictable performance.

Hardware emulation is the basis of system verification offering combined high verification speed and confidence on the achieved results. Tests can be performed with real-world I/O if the rest of the system exists, test vectors can be fed from mass storage devices, or test benches can be compiled into the emulator to achieve comprehensive verification. Similarly, performance and functionality verification can be implemented on the emulator. In addition, these methods facilitate data collection over large sample sets, which are needed to validate bit error rates in high signal to noise ratio environments where the error rates can be extremely low. This tight coupling of the hardware emulation even in the early stages of the design and its implementation facilitate the primary goal of early performance evaluation.

To demonstrate the approach and method of description, a simple 1 Mbit/s transmission system was mapped to 3 FPGAs and run at 32 MHz, thus achieving real-time behavior in emulation conditions. The same description was then used to generate an ASIC layout that functionally had exactly the same performance, but used a logical representation more optimized for standard cell implementation. Advantages of this method, from the designer's point of view, include improved understanding of the overall system and its real-time behavior with the analog portions of the system, effectively eliminating the simulation speed bottlenecks, automatic test bench generation, and interoperability with other analysis software such as the Matlab tools.

ACKNOWLEDGEMENTS

Dr. Kuusilinna's work was supported by the Technology Development Center of Finland (Tekes), Jenny and Antti Wihuri Foundation, and the Finnish Cultural Foundation.

This work was supported by DARPA and MARCO under the Center for Circuits, Systems and Software (C2S2) in the Focus Centers Research Program and the MURI program sponsored by the U.S. Army Research Office. In addition, we would like to thank Xilinx for donating the FPGA chips. Finally, we would like to acknowledge the support of the members of the Berkeley Wireless Research Center.

REFERENCES

1. J. Babb, R. Tessier, and A. Agarwal, "Virtual Wires: Overcoming Pin Limitations in FPGA-based Logic Emulators," *Proc. IEEE Workshop on FPGAs for Custom Computing Machines*, pp. 142-151, Apr. 5-7, 1993.

2. M. Butts, J. Batcheller, and J. Varghese, "An Efficient Logic Emulation System," *Proc. 1992 IEEE Int'l Conf. Computer Design: VLSI in Computers and Processors*, pp. 138-141, Oct. 11-14, 1992.

3. M. Courtoy, "Rapid Prototyping for Communications Design Validation," *Conference Record Southcon/96*, pp. 49–54, June 25-27, 1996.

4. W.R. Davis, *et al.*, "A Design Environment for High-Throughput, Low-Power Dedicated Signal Processing Systems," *IEEE J. Solid-State Circuits*, Vol. 37, No. 3, Mar. 2002.

5. H. Krupnova, Dinh Duc Anh Vu, G. Saucier, and M. Boubal, "Real Time Prototyping Method and a Case Study," *Proc. 1998 9th Int'l Workshop on Rapid System Prototyping*, pp. 13-18, June 3-5, 1998.

6. H. Krupnova, C. Rabedaoro, and G. Saucier, "FPGA Partitioning for Rapid Prototyping: A 1 Million Gate Design Case Study," *Proc. 1999 IEEE Int'l Workshop on Rapid System Prototyping*, pp. 13-18, June 16-18, 1999.

7. A.D. Pimentel, *et al.*, "Exploring Embedded-Systems Architectures with Artemis," *Computer*, Vol. 34, No. 11, pp. 57-63, Nov. 2001.

8. F. Slomka, M. Dorfel, R. Munzenberger, and R. Hofmann, "Hardware/Software codesign and Rapid Prototyping of Embedded Systems," *IEEE Design & Test of Computers*, Vol. 17, No. 2, pp. 28-38, Apr.-June 2000.

9. B. Spitzer, M. Kuhl, and K.D. Muller-Glaser, "A Methodology for Architecture-Oriented Rapid Prototyping," *Proc. 12th Int'l Workshop on Rapid System Prototyping*, pp. 200-205, June 25-27, 2001.

10. www.mathworks.com.

11. www.xilinx.com.

Chapter 11

TECHNOLOGY CHALLENGES FOR SOC DESIGN
An IBM Perspective

John M. Cohn
IBM Corp.

Abstract: Advances in silicon technology have fueled much of the growth in SoC performance and density. Yet these technology advances have not come easily: each new technology generation brings with it a new set of technology related design considerations and challenges. This chapter outlines the major technology issues facing current and future SoC designers. These challenges are divided into issues that influence design, those that influence manufacture and those that relate to affordability. The design issue section outlines problems relating to performance scaling, power management, signal integrity, parametric variability and reliability. The manufacturability section describes issues relating to manufacturing yield and lithography challenges. Finally, the affordability section covers issues relating to rising engineering costs and technology trends. Each topic begins with a summary of the principal issues and trends, followed by a discussion of the associated SoC impact and mitigation strategies.

Key words: Performance scaling, power dissipation, signal integrity, variability, reliability, lithography scaling, yield, engineering cost, technology integration challenges.

1.   INTRODUCTION

Modern SoCs are marvels of engineering design. They combine in a single silicon chip what once required a roomful of equipment. Advances in silicon technology have fueled much of the growth in SoC performance and density. Yet, as any SoC designer will tell you, these technology advances have not come easily. Each new technology generation brings with it a new set of technology related design considerations and challenges. This chapter outlines the major technology issues facing current and future SoC designers. In the interest of clarity, these challenges are divided into issues that influence design, those that influence manufacture and those that relate to affordability. The design issue section outlines the problems relating to performance scaling, power management, signal integrity, parametric variability and reliability. The manufacturability section describes issues relating to manufacturing yield and lithography challenges. Finally, the affordability section covers issues relating to rising engineering costs and technology trends. Each topic begins with a summary of the principal issues and trends, followed by a discussion of the associated SoC impact and mitigation strategies.

2.   DESIGN ISSUES

This section outlines the main technology-related design issues facing a modern SoC designer. To understand these issues, it is first important to appreciate the advances in performance and density that have characterized the last decade of SoC development. Innovations in technology have allowed device performance to double every eighteen months and chip density to double every two years in keeping with the ever-quoted 'Moore's Law' rate. Figure 11-1 shows this growth in terms of the device delay and chip density of IBM ASIC/SoC technologies over the past four generations [1].

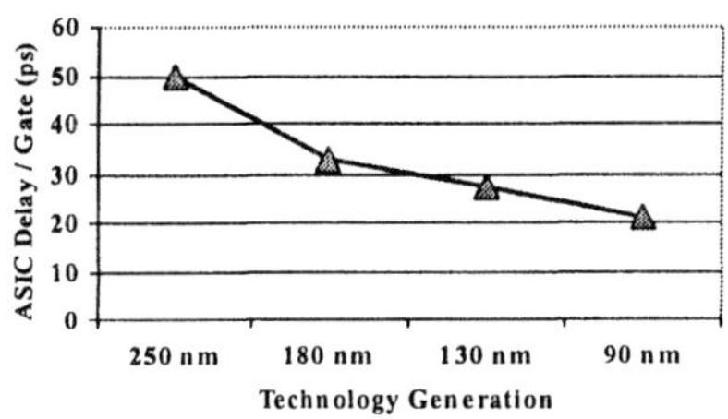

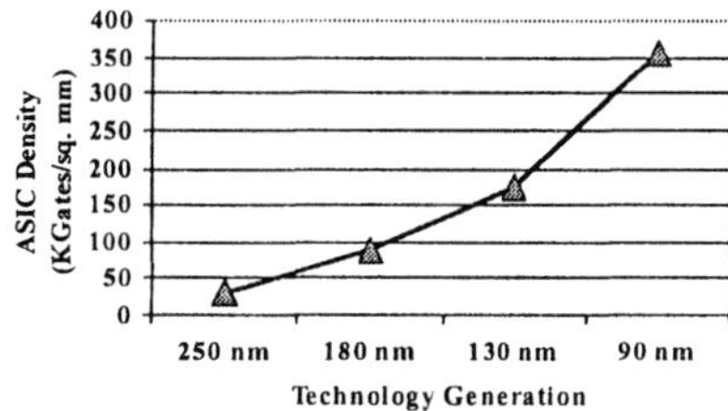

Figure 11-1. SoC Performance and Density Growth

This quest for increased performance and density has given rise to a number of very complex design-related problems which complicate SoC design. This section outlines several of these challenges, including those of performance scaling, power dissipation, signal integrity, parametric variability and reliability.

2.1   Performance scaling

One of the assumptions of SoC design has always been that each new technology brings with it a boost of 15 to 20 percent in system-level performance. The industry is now entering a phase where those performance gains are becoming increasingly difficult with each new process generation. In the early days of VLSI, most of the performance improvement came from simple geometric scaling of device and interconnect dimensions. As device dimensions began to shrink below 0.25 microns, however, it no longer proved possible to scale all device parameters of threshold voltage, gate oxide thickness, gate length and supply voltage at the same rate. For example, it was no longer possible to continue to reduce threshold voltages at the historical rate due to device leakage considerations. As this happened, the performance impact of scaling began to fall off. To make up for this performance shortfall, process designers began incorporating more sophisticated materials and processing to compensate for the shortfall in both device and interconnect performance.

2.1.1   Technology Trends

IBM has been a leader in the process innovations which have helped keep performance scaling on track. For example, IBM was first to replace aluminum wiring with lower resistance copper wiring in its microprocessor and ASIC technologies in the 0.18 micron technology node. This lower resistance has lead to a roughly 30 percent reduction in wiring RC delay [2]. At the 0.13 micron technology node IBM has been a leader in the introduction of low permittivity interlayer dielectric materials. These new materials lower wiring capacitance which further reduces RC delay by 15 percent [3].

As the industry scales to below 0.13 microns, process designers must resort to even more sophisticated materials to keep performance scaling on track. At IBM, technology improvements already pioneered for the company's industry leading microprocessor products will be applied to SoC technology. One such example is IBM's *silicon on insulator (SOI)* technology in which devices are grown on insulating silicon dioxide regions to reduce device parasitics. SOI technology has already been used in IBM's industry leading microprocessor technology to boost circuit performance by as much as 20 percent [4]. IBM researchers have also demonstrated the advantages of *strained silicon,* a process in which device structures are grown on germanium doped silicon regions which induce oriented strain in a device. The resulting improvement in channel carrier mobility has been shown to improve device performance again by as much as 20 percent [5].

Other near-horizon improvements that will likely be applied to SoC design include high dielectric (*high K*) gate materials such as *hafnium dioxide* to allow improved gate drive at lower gate leakage, and lower permittivity *porous low-k* inter-level dielectrics for further interconnect capacitance reduction. In addition to these material changes, device structure changes such as vertical dual gate *'FinFets'* have already been shown to almost double the device transconductance [6]. Figure 11-2 illustrates some of these future process options.

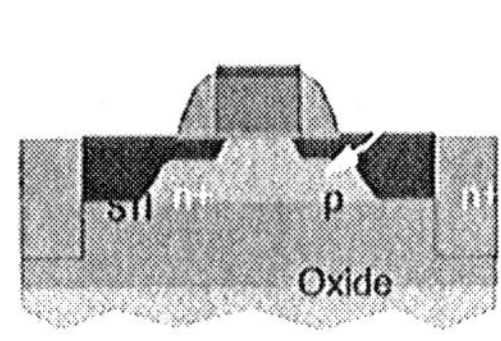

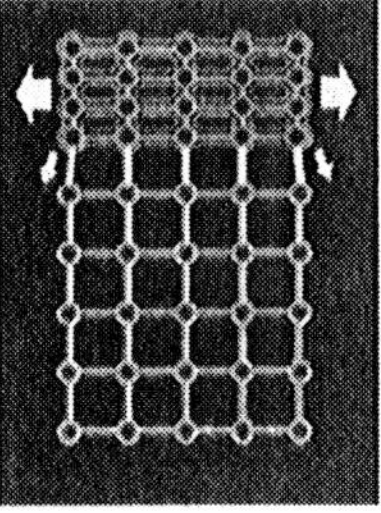

Copper with Low K   **Silicon On Insulator**   **Strained Silicon**

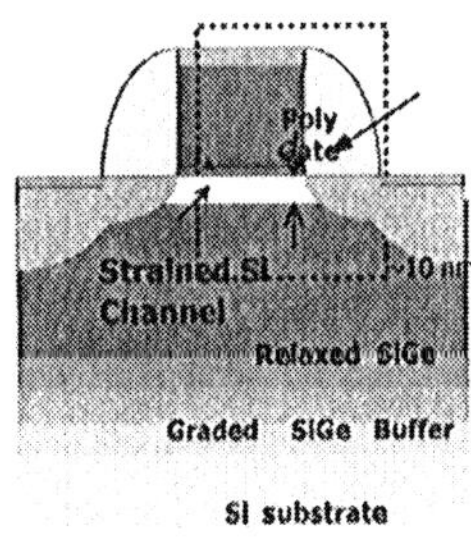

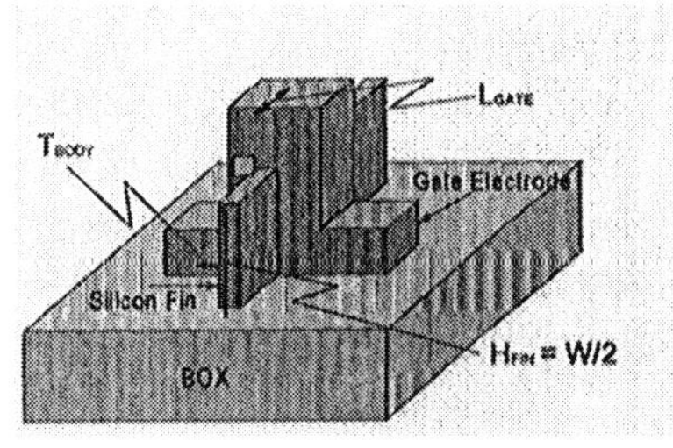

High K Gate Material   **Fin FET**

Figure 11-2. IBM Copper Interconnect and Silicon on Insulator Technology

The increasing process complexity required to maintain performance scaling unfortunately has its own direct impact on the design and reuse of SoCs and their constituent IP. Each new technology feature introduced in the pursuit of performance introduces new complexity in the electrical modeling, layout, analysis and design methodology of VLSI circuitry. One powerful example of this can be seen in the introduction of SOI technology for IBM's advanced servers. SOI has given IBM a significant performance advantage in very high end microprocessors. At the same time, the introduction of SOI introduced the need to model and manage a host of new electrical effects including body voltage hysteresis, parasitic bipolar currents and device self heating [4].

In addition, the introduction of new technology features can impact the ability to migrate SoC IP from one technology generation to the next. For example, the addition of a new feature such as SOI can require significant redesign of existing *hard* IP such as arrays, IO blocks and microprocessor cores.

The best way to minimize the reusability impact of process evolution on design IP is to capture the design in the most abstract form possible. For example, it can be very difficult to remap an IP block such as a microprocessor core if it exists only as a *hard* macro in GDSII. If instead the microprocessor design is captured as high-level RTL, it can be remapped relatively easily to a next generation technology using logic synthesis.

2.1.2   Interconnect performance gap

Even with all of the above mentioned process innovations, it will likely be impossible to keep device and interconnect scaling in lock-step. Improvements in interconnect performance have begun to lag improvements in device performance which has lead to an interconnect performance 'gap'. Figure 11-3 illustrates this divergence. While gate delay has decreased considerably in the technology generations between 0.50 micron and 0.13 micron, the delay of a fixed length of interconnect has increased. This increase is due partially to the increase in resistance caused by shrinking wire cross sections and partially to an increase in lateral capacitance due to smaller inter-wire spacing.

To understand the impact of the interconnect performance gap, it is helpful to think of interconnect as divided into two types: *local* and *global*. Local interconnect connects the transistors in a small region such as within an IP block. Global interconnect is used to interconnect IP across large distances on the chip. As device density increases, local interconnect lengths can also be reduced. For this reason local interconnect tends to scale relatively well. Global interconnect, by contrast, is not scaling well. With chip sizes remaining roughly constant, the relative delays of global interconnect are increasing [7]. The impact of this divergence is profound. Where once it was possible to assume that all IP on a chip was within a single clock cycle, it may now require multiple clock cycles for a critical signal to travel across the die.

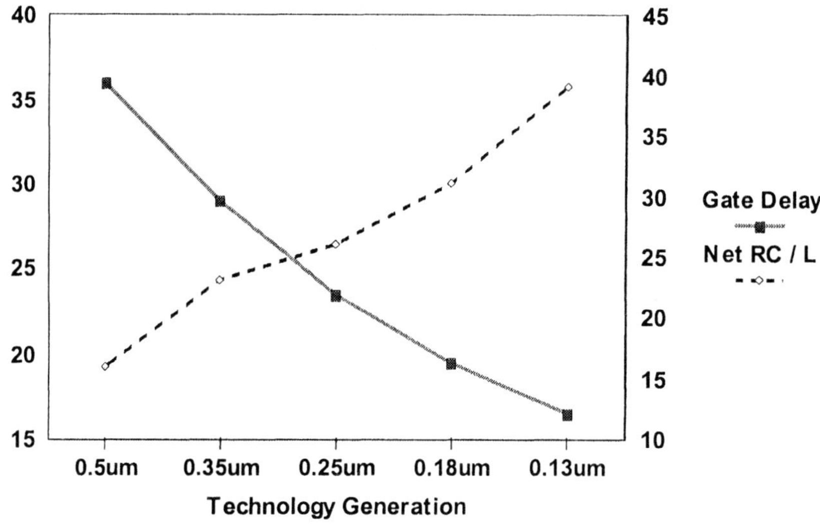

Figure 11-3. Device and interconnect performance scaling

To account for the possibility of multi-cycle global interconnect, it may now be necessary to accommodate interconnect delays at an architectural level using explicit pipelining. In a bus-based SoC this may require the splitting of a single logical bus into multiple local regions or the addition of pipelining into emerging bus specifications [8].

In the future it may prove worthwhile to compose SoCs from locally iso-synchronous units which communicate across large on-chip distances using asynchronous signaling. Eventually, even more sophisticated packet-based *network-on-chip* protocols may prove useful for global communications on large future SoCs [9].

2.1.3   Diagonal routing

One intermediate term solution to poor global wire length scaling in SoCs is the introduction of 45 degree or diagonal routing. For simple two point connections, diagonal routing can be shown to reduce wire length by a factor of $\sqrt{2}/2$. In the more general case of multi-point nets, average wire length reductions of approximately 20 percent have been observed [10]. In addition to this average wire length reduction, diagonal routing has been shown to reduce via counts. The combination of shorter wires and fewer vias can reduce wire delay by as much as 30 percent while at the same time enabling smaller die sizes and improved yield for future SoCs.

While the move to allow diagonal routing could improve many design parameters in future SoCs, its introduction will not be easy. In the past, SoC wiring has been restricted to orthogonal routing only. This orthogonal assumption has been used to simplify many tool algorithms in current SoC integration flows. These include the routing algorithms themselves as well as associated issues such as wire length prediction, geometry extraction and ground-rule checking. Many tools in the current SoC integration flow will have to be enhanced to enable efficient handing of diagonal routing before this technique can be used pervasively. A very interesting proof of concept demonstration of this flow has already been made by Simplex, now part of Cadence Design Systems [11].

2.2 Power dissipation

Power dissipation is emerging as one of the fundamental limits to SoC complexity scaling. Until recently power was of significant concern only to those designing at the extremes of the power spectrum: those designing ultra-low power SoCs for battery powered applications, or those designing ultra-high performance SoCs for applications such as super computers. Increasingly, however, power density fueled by huge growth in circuits-per-chip is threatening to make power dissipation a limiting design constraint for SoCs of almost all applications.

There are two types of power management issue facing SoC designers today: active power and leakage power. Both are increasing at alarming rates. To understand this power crunch it is important to understand the mechanisms behind both types of power dissipation. Active power is dissipated through the charging and discharging of the capacitance of the switching nodes. The magnitude of this power is given by Equation 1.

$$P_{Active} \propto \frac{1}{2} C * Vdd^2 * F \qquad (1)$$

Where C is the total switching capacitance, V_{dd} is the supply voltage and F is the switching frequency. As silicon technology scales, the capacitance per unit area and the frequency of operation increase by 30% for each technology generation. Assuming perfect scaling, these increases are exactly offset by a corresponding 30% decrease in V_{dd}, and the power per unit area remains constant. Unfortunately, frequency of operation has increased at a faster rate than the scaling of the silicon process technology [12]. This has led to an increase in active power density in each technology generation as shown in Figure 11-6. Figure 11-4 shows the most recent ITRS projection of supply voltage and total power [13]. Note the steady increase in supply current and total power despite the aggressive drop in supply voltage.

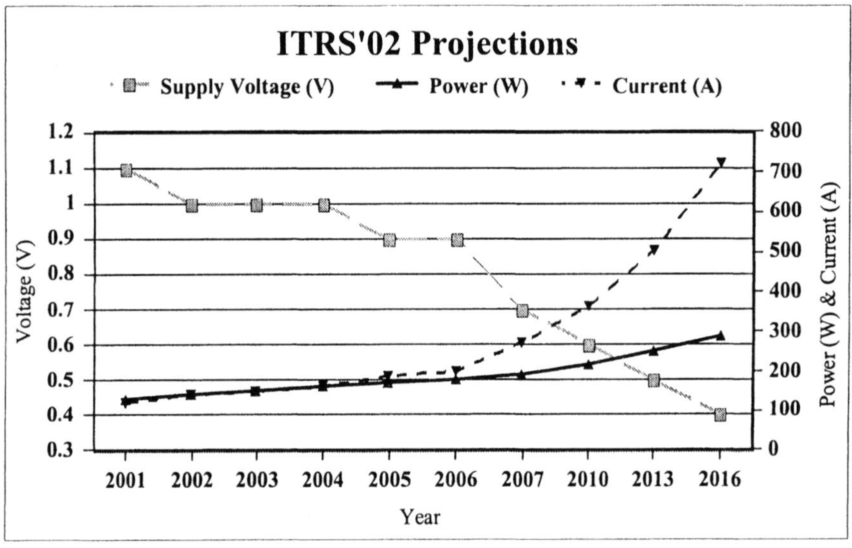

Figure 11-4. SoC supply voltage and power trends

The second component of power dissipation is *leakage power* which is caused by the current that leaks through a device even when is turned off. Leakage power is increasing due to several aspects of device scaling. As silicon technologies advance, smaller geometries become possible, requiring improvements of device structures including lower transistor oxide thickness (T_{ox}), which in turn increases transistor performance. To maintain circuit reliability, the supply voltage (V_{dd}) must be lowered as T_{ox} is reduced. As V_{dd} is reduced, the transistor threshold voltage (V_t) must also be reduced in order to maintain circuit performance. This decrease in V_t then drives significant increases in the amount of in *sub-threshold* current that can leak through a device which is switched off. Another consequence of thinning gate insulators is *gate leakage*. Beginning in the 90 nm technology generation, tunneling currents through the gate of the device will also become a non-negligible component of leakage power. Figure 11-5 shows these future gate leakage trends in the 1 – 4 nanometer range. The figure also includes an atomic force micrograph showing the 2- 3 nanometer gate oxide interfaces that will soon be required by scaling.

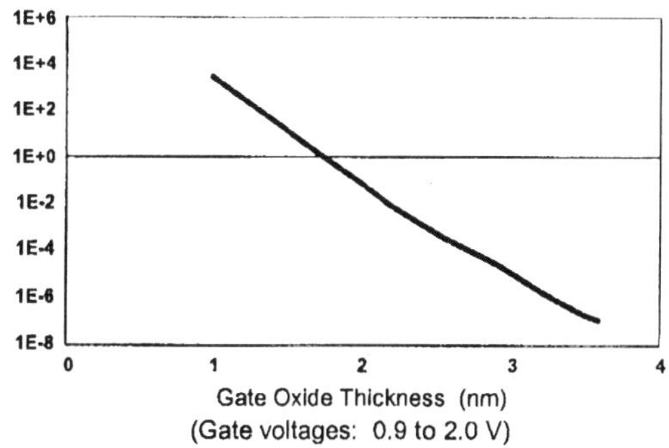

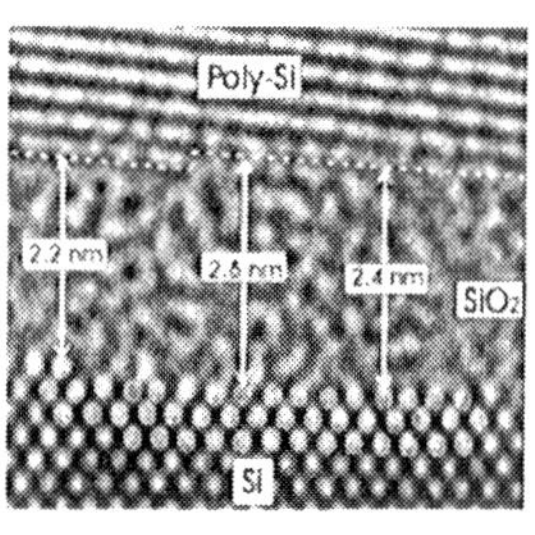

Atomic force microscope
image of gate oxide interface

Figure 11-5. Gate leakage trends

The increase in both active and leakage power is evident in Figure 11-6 which shows the power density of numerous ASIC/SoC designs plotted against technology generation. Note that at the current rate, leakage power density will approach that of active power in roughly the 0.045 micron (45 nm) technology node.

The increase in both types of power has a significant impact on system cost in the areas of chip packaging, system cooling, power supply design and test.

Chip package costs can be a significant portion of finished product cost for an SoC. The higher the power dissipation a chip has, the more expensive the package required. For example, an SoC that might be electrically compatible with an inexpensive plastic package may be forced to use a much more expensive ceramic package because the more expensive package has superior heat dissipation properties. Similarly system cost, space and environmental noise increase dramatically when high powered SoCs force the use of large heat sinks, cooling fans, chilled air and/or circulating fluid refrigeration.

In the case of hand-held devices both active and leakage power is the ultimate limit of battery life. The higher the power of the SoC, the larger, heavier and more costly its battery must be. This trend is actually worsening as SoC power is increasing at a faster rate than battery technology is improving [14].

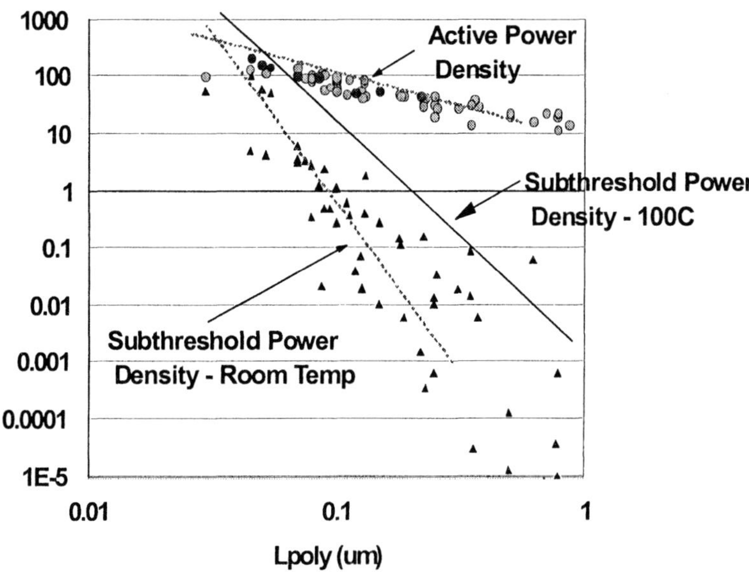

Figure 11-6. Power dissipation trends

Finally, increased power also puts pressure on manufacturing test. Increased leakage current levels are interfering with the ability to use quiescent current or *IDDq* measurements as a means to identify faulty chips. Higher supply currents are also taxing *burn-in* equipment that test chips while subjecting them to extended periods of elevated temperature and voltage stress to weed out marginal products, particularly as leakage current increases rapidly with temperature. Higher SoC supply currents are beginning to limit the number of chips that can burned-in at the same time in a given oven. The result is longer test times and ultimately higher product cost.

2.2.1   Power prediction for SoCs

Perhaps the most important aspect of designing for lower power is the need to accurately predict the power dissipation of a design before building it. It may come as a surprise to some that this remains a difficult problem for complex SoCs. The difficulty of power prediction is rooted in the same complexity that makes power such a problem. For a very small circuit it may be possible to use detailed circuit simulation to accurately calculate both active and leakage power over *all* possible operating states. However, on a large SoC with tens to hundreds of millions of switching elements, it is difficult to simulate even a single operating state, much less consider simulating the extremely large number of possible operating states of all elements on the chip. As a result, researchers have devised a number of static analysis methods that attempt to calculate approximate switching power without resorting to simulation. These methods may combine statistical estimates on switching activity on inputs with static timing information to place better bounds on estimated switching activity to improve accuracy [15]. Even so, these methods remain quite inaccurate, with correlation to power measured on hardware often off by as much as 30 %.

Despite this difficulty, it is becoming increasingly important to get good bounds on power at the earliest phases of design to allow architectural tradeoffs, to make correct package selection and to understand system cooling requirements. Improving the quality of these early power predictions is currently an area of intense research in the industry.

2.2.2   Power Optimization in SoCs

Once SoC power can be accurately predicted, it is possible to apply a number of technology, circuit and architecture techniques to optimize for lower active and leakage power. Some of the most effective techniques include clock gating, voltage scaling, multi-threshold logic, and the use of voltage islands.

2.2.2.1   Clock gating

As clock speeds increase, and skew tolerance decreases, an increasing percentage of chip power is dissipated in the distribution of the system clock. In a simple synchronous design every portion of the chip is switching in response to the clock even if every portion is not doing useful work in that clock cycle. Significant power reductions can be attained through *clock gating*, [16] the act of stopping the clock in sections of the chip that are not being used. By blocking logic switching in a logical region of the chip, clock gating can significantly reduce active power. The introduction of clock gating does introduce additional complexity in the definition of on-chip clock domains and their timing. In particular, SoC IP which makes use of clock gating must take special care to ensure that the timing of the logic signals that control gating of the clock are constrained in order that they do not interfere with proper logical function of the chip.

2.2.2.2   Voltage scaling

Another powerful technique which can be used to combat power is voltage scaling. Because active power is proportional to the square of the supply voltage, a small reduction in supply voltage can lead to a significant reduction in active power. Lowering supply voltage also reduces leakage power by decreasing sub-threshold current. Unfortunately, this power saving comes at the expense of performance as lowering supply voltage also increases the delay through switching. The design objective in voltage scaling then becomes to find the lowest supply voltage that allows the chip to run at the desired performance [17].

2.2.2.3   Multi-threshold logic

The use of multi-threshold libraries is becoming a common method for better handling of this trade-off of power and performance. Increasingly, technologies provide a range of discrete threshold options in their device menu. Typically this includes a standard *Vt*, a low *Vt*, and a high *Vt* option. These devices are used to create ASIC/SoC libraries with different performance and power characteristics. Figure 11-7 illustrates the relationship between delay and supply voltage for regular and low Vt logic gates in 130 nm technology.

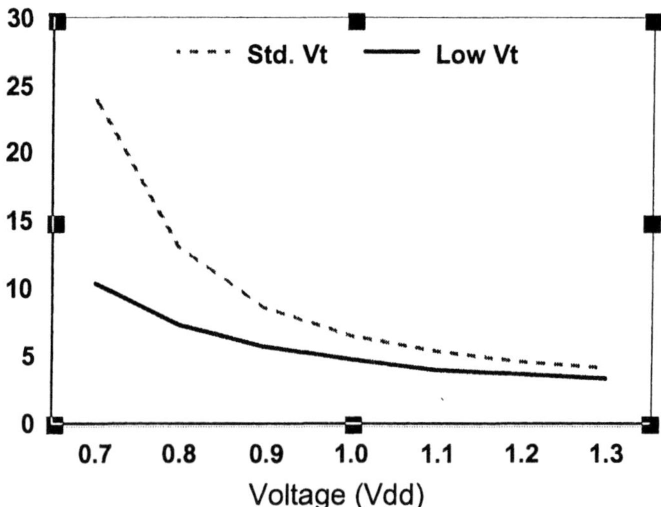

Figure 11-7. Mixed threshold logic for power reduction

As can be seen, circuits using low *Vt* devices provide a performance advantage over higher threshold transistors, particularly at lower voltage. Using low *Vt* circuits can allow timing closure at a lower voltage level, which can be a great saving for overall active power. Unfortunately, low device thresholds also imply higher levels of leakage current. For this reason, most applications must limit their use of low *Vt* logic to the most timing critical portions of the design.

In contrast, high *Vt* logic is designed to have lower leakage power than either standard or low *Vt* logic, at the expense of lower performance. For applications requiring very low standby leakage current, logic elements utilizing high *Vt* logic can be placed in non-timing critical portions of the design. The design objective in multi-threshold optimization then becomes to find the mix of normal, low and high threshold circuits that best balances performance, active power and leakage power. Clearly, this is a very difficult process to optimize by hand. IBM has automated the process as part of its *placement driven synthesis (PDS)* flow [18]. PDS uses its integrated timing analysis to identify critical paths which are candidates for low threshold logic. The substitution is made subject to a calculation of total leakage current such that the total chip leakage budget is not exceeded. A similar substitution can be made on paths with large positive slack. In these cases PDS can selectively substitute high threshold logic to trade timing slack for lower leakage.

2.2.2.4 Voltage islands

Recently IBM has implemented a method called *voltage-islands* [19] which combines clock gating, voltage scaling and mixed threshold optimization. In its simplest form, voltage islands is a technique which manages power by allowing different sections of the chip to be run at different operating voltages. Sections which must run fastest are powered by higher voltages, while those that can run slower are powered by lower voltages. For example, performance-critical elements such as a processor core may require the highest voltage level supported by the technology in order to maximize its performance. Other less performance critical elements such as memories or control logic may not require this level of voltage, saving significant active power if they can be run at lower voltages.

The voltage island concept can be combined with clock gating and mixed threshold logic to further optimize total chip power and performance. There are several variants of this voltage island concept. In the simplest scenario, sections of the chip are run off different fixed voltage supplies. In other cases, the voltage can be dynamically adjusted for one or more islands to adapt to changing performance requirements. In still other cases, the voltage may be completely removed from inactive regions to further lower leakage power. Commonly, complex SoC designs consist of a number of diverse functions, few of which are active at any given time. By shutting off power to inactive islands both active and leakage components of power can be significantly reduced. The voltage island concept is illustrated below in Figure 11-8. The figure shows two main voltage islands powered by separate voltages *V1* and *V2*. Each island contains a single sub-island powered by voltages *V11* and *V21* respectively. The power to each island is controlled by a power switch that allows the island to be powered down. The power sequencing for islands *V1* and *V2* are controlled by a central power management unit, while the power sequencing to islands *V11* and *V21* are controlled by internal logic.

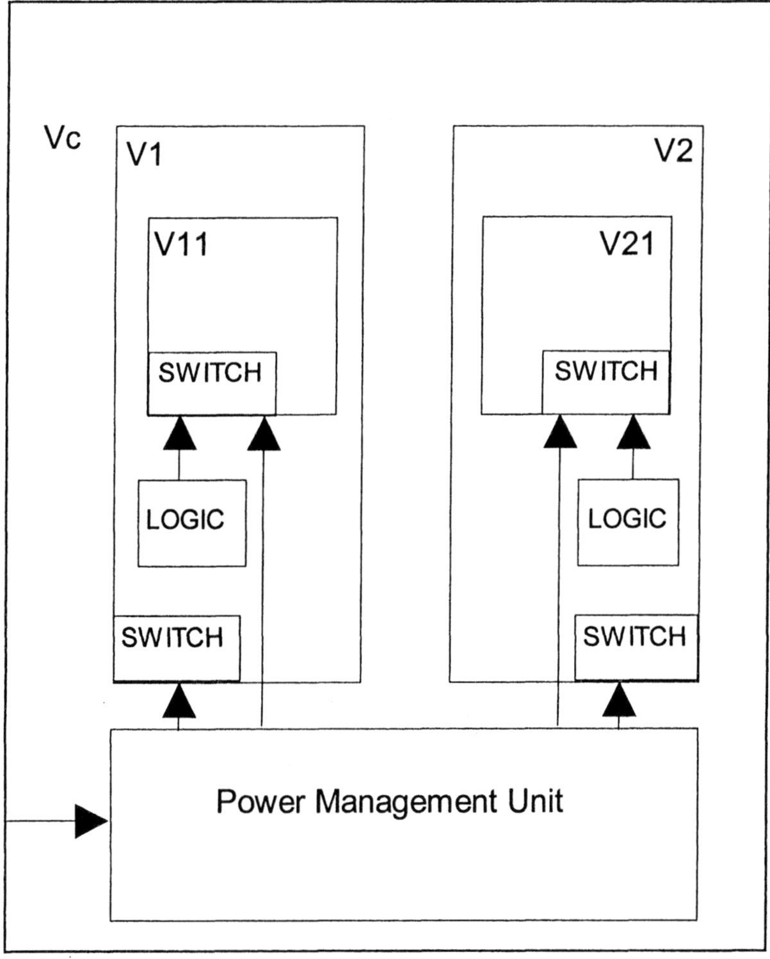

Figure 11-8. Voltage Islands

One example where voltage islands were effective at limiting power was a 90nm chip which used 144 high speed serial links, 10M logic gates and 20M bits of SRAM. The links ran at a speed of 5 GHz, while the rest of the chip ran at a more modest 500 MHz. In order to run at the required performance, the links required an operating voltage of 1.2V. The active power of the original design was estimated at 36.6W with a standby power estimated at 0.5W. By placing the links and their associated high speed logic in one Voltage Island, and separating the rest of the logic into its own voltage island, the power was reduced to 30.7W and the standby power was reduced to 0.25W. In this scenario, the links were supplied a continuous 1.2V supply while the rest of the logic utilized a 1.0V supply.

2.3　　　Signal integrity

In addition to performance and power, SoC designers are faced with a growing list of signal integrity or noise concerns. These concerns are heightened as supply voltages scale down and noise margins erode. The dominant noise concerns in a modern SoC are coupling noise, power supply noise and substrate noise.

2.3.1　　　Coupling noise:　Issues and Trends

Coupling noise occurs when a voltage transition on an aggressor wire causes current to be injected into an adjacent *victim* wire through the two wires' mutual capacitance.

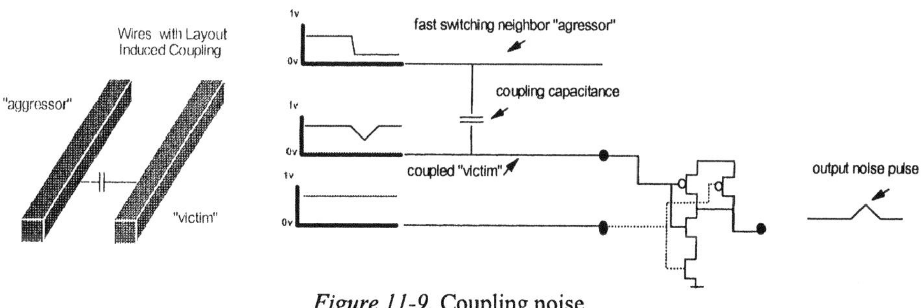

Figure 11-9. Coupling noise

This injected current can then give rise to a voltage change on the victim wire. This relationship is show in Equation 2 below.

$$V_{coupling} \propto Z_{drive} \frac{C_{coupling}}{C_{total}} \frac{dV}{dt} \qquad (2)$$

Where $V_{coupling}$ is the voltage change in the victim wire, Z_{drive} is the wire's driving impedance, $C_{coupling}$ is the mutual capacitance shared by the aggressor and victim wires, C_{total} is the total capacitance on the wire including coupling, self capacitance and gate loading and dV/dT is the voltage rate of change between the aggressor and victim.

If the coupled signal exceeds the logic threshold on the victim wire, an incorrect logic value or *glitch* will be induced in the victim circuit. If the glitch propagates into a latch during its set-up, the glitched data can be captured by the latch causing a functional fail. Glitch errors are only one type of coupling problem. If the victim wire is transitioning during the coupling event, its transition time can be significantly affected. If the aggressor and victim wire are transitioning in the opposite direction, e.g. one rising, one falling, the transition on the victim wire can be significantly delayed. Alternately, if the transitions on the aggressor and victim wires are in the same direction, e.g. both rising, the transition on the victim wire can be significantly sped up. If not modeled correctly, this variation in delay can cause unexpected variations on chip performance. If these delay variations are large enough they can lead to improper operation.

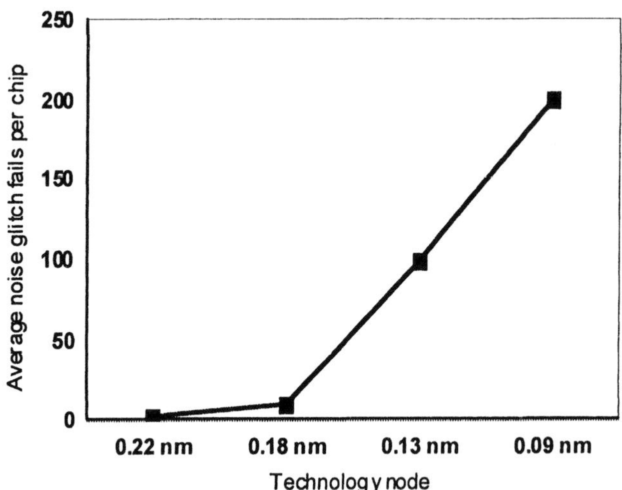

Figure 11-10. Analyzed noise glitch fails per chip

Coupling noise analysis in SoCs has changed markedly since the 0.25 micron technology era. In those days, coupling noise issues were rare and quite predictable. It was generally sufficient to use simple rules of thumb to check sensitive signals such as long un-buffered wires. Once identified, these signals could be manually modified to reduce their noise sensitivity. This situation has changed markedly. As interconnect pitches have shrunk, and scaled wire lengths, drive impedances and switching frequencies have increased, the number of wires susceptible to significant delay changes and noise glitches is increasing. This increased sensitivity to noise coupled with the ever increasing number of wires on a chip has resulted in an almost exponential increase in the number of noise issues detected on chip. Figure 11-10 shows this large rise in noise problems found and fixed during noise analysis in IBM design centers.

2.3.1.1 Coupling noise mitigation

Once coupling problems are identified, there are many techniques which can be used to repair them. The simplest fix is often buffer insertion which can be used to lower total coupling capacitance on any single wire segment. In addition, driver sizes can be increased to reduce driving point impedance which improves noise tolerance on shorter wires. It is also possible to reduce coupling by increasing the wire to wire spacing between aggressor and victim wires. For the most noise sensitive wires, it is even possible to insert grounded *shield nets* around them to prevent nearly all coupling.

As the number of signals identified as coupling problems increases, it is becoming less practical to fix them manually at the end of the layout process. To avoid this becoming a major productivity and design quality bottleneck, it is necessary to move from a design flow of noise analysis and fix-up to one of noise avoidance. There are several techniques which can be used to prevent noise problems from being introduced during design. IBM's Placement Driven Synthesis (PDS) flow uses gate sizing and wire buffering rules to prevent long, under-driven wires from occurring in the design. IBM's XROUTER global/detail router can permute wires during global wiring to avoid many potential coupling problems. IBM's ET-coupling tool then analyzes the effect of the remaining coupling on timing. Finally IBM's Routing Based Optimization (RBO) can be used following final routing to identify, re-route, resize and/or re-buffer the few coupling noise problems that remain.

As SoCs become more complex, it will be increasingly important to be able to predict noise earlier in the design flow. This will require a more interconnect-centric view of chip floor planning. Emerging *virtual prototyping* tools will be useful to get better predictions of wire lengths and noise sensitivities even at the RTL planning level.

2.3.2   Power supply noise

Power supply noise is emerging as one of the most difficult design problems to predict in an SoC. Power supply noise is the result of voltage drops in the power supply routing due to the finite impedance of the supply routing and the time varying current needs of the chip logic. As a chip logic element switches, current must be sourced from off chip through the chip package as shown in Figure 11-11.

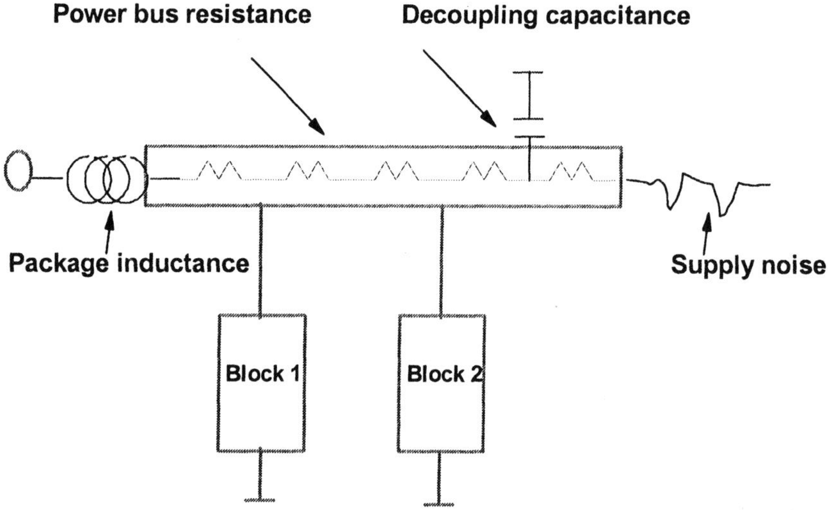

Figure 11-11. Power supply coupling noise

As this current flows, it encounters both the resistance and inductance of the power supply. This induces a time varying voltage drop at every point along a path between the switching element and the off chip power supply as shown in Equation 3 below.

$$V_{drop} \propto L_{pkg/chip} \frac{di_{switching}}{dt} + R_{pkg/chip} i_{switching} \qquad (3)$$

Where $L_{pkg.chip}$ is the total chip and package inductance, $R_{pkg/chip}$ is the total resistance of the package and chip power routing, $i_{switching}$ is the switching current and $d\,i_{switching}/dt$ is the time rate of change of the switching current.

The trend in power supply noise is alarming. Increases in switching speed combined with increases in total chip current give rise to larger $L\,di/dt$ and resistive drops. At the same time the chip voltage supply is decreasing which make these voltage variations proportionally even more damaging.

When added together, the voltage drops induced by each switching event leads to potentially large variations in supply voltage both in space and in time. Figure 11-12 shows a map of the spatial distribution of the average or steady-state voltage drop across a single large ASIC calculated by IBM's ALSIM tool [20]. The areas of similar fill density correspond to areas of similar voltage drop. This plot shows a range of 1.2V in the chip corners all the way down to 0.9V in the chip center. The time domain variations in voltage can be equally disruptive. In this same example the supply voltage measured at one point on the Vdd rail rises as high as 1.3 V and falls as low as 0.9 volts before settling around the nominal supply voltage of 1.2V. The large overshoot and undershoot are due to the *di/dt* of the switching current being supplied through package and chip inductance.

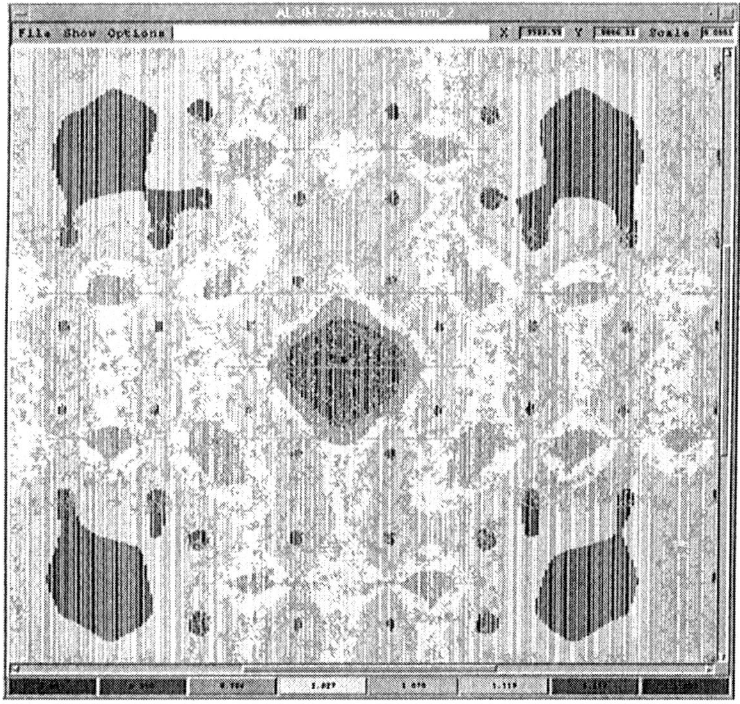

Figure 11-12. Supply drop voltage map

These spatial and time varying voltage variations, in turn, affect the performance of every other logic circuit on the chip. To illustrate this effect Figure 11-13 shows supply induced delay variation in a simple 90 nm 2-way AND. A supply voltage reduction of only 100 mV from the nominal 0.9 V supply is seen to degrade the delay of the gate by 27 percent.

When one thinks about the combination of millions of logic elements switching in a given clock cycle, the complexity becomes staggering. Just as in power dissipation analysis, it is very difficult to accurately predict the detailed switching characteristics of every logic element on a chip. Currently the industry uses a combination of statistical methods to estimate switching activity, to be then used to calculate power supply drop. At present no viable solution exists that models the effects of all possible supply collapse scenarios back into performance analysis.

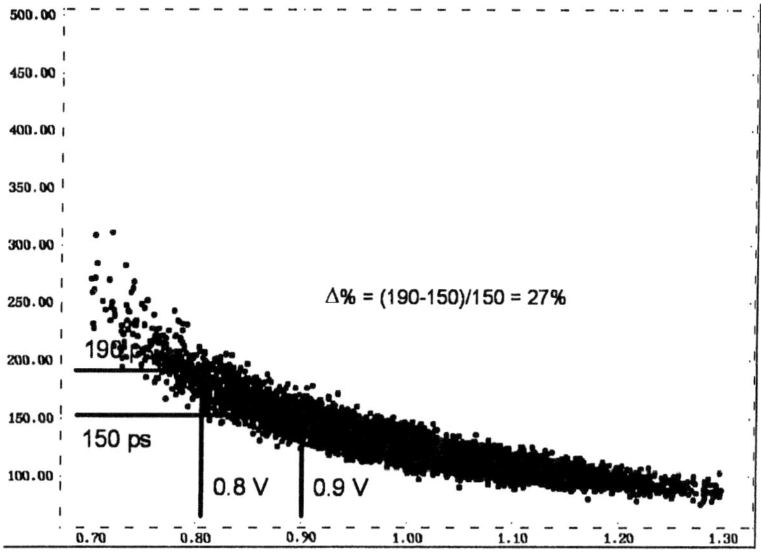

Figure 11-13. Supply voltage induced delay variation

One standard way to minimize the impact of power supply noise is to create more robust power supply routing. This entails devoting more routing resource to power distribution. In the future it may be necessary to devote even larger portions of entire routing layers to power distribution. In addition to making the package and chip power routing more robust it is also possible to reduce the effect of package inductance though improvements in package materials and the application of careful layout techniques including designing current return paths for inductive loops. It can also be helpful to isolate the power supply routing of noisy and sensitive portions of the chip.

This technique of power supply isolation is particularly important in mixed-signal SoCs.

The final weapon in combating power supply droop is the careful placement of decoupling capacitors. Decoupling capacitors act as small charge reservoirs that are placed throughout the chip to help source the instantaneous current requirements of a switching circuit. This may require accommodation in floor planning and placement of an SoC to ensure that there is sufficient space reserved for the required amount of decoupling. It is also important, however, to use no more decoupling capacitance than is necessary, as the gate structure in decoupling devices contributes to total chip leakage power.

Figure 11-14 illustrates a power supply noise problem detected on a 180nm SoC. In the original design, a group of high power SRAM arrays were placed very close to a voltage sensitive high speed serial link in one corner of the chip. Analysis with IBM's ALSIM program revealed that the switching current from the SRAMs created more than 160 mV of power supply droop in the supply to that corner of the chip. This supply degradation was well outside the allowable performance window for the serial link. To correct the problem, the SRAMs were moved further away from the serial link, and 15nF of additional decoupling capacitors were added to decouple the sensitive serial links. Analysis revealed that this mitigation kept the supply droop well within the 100mV operating range of the serial link.

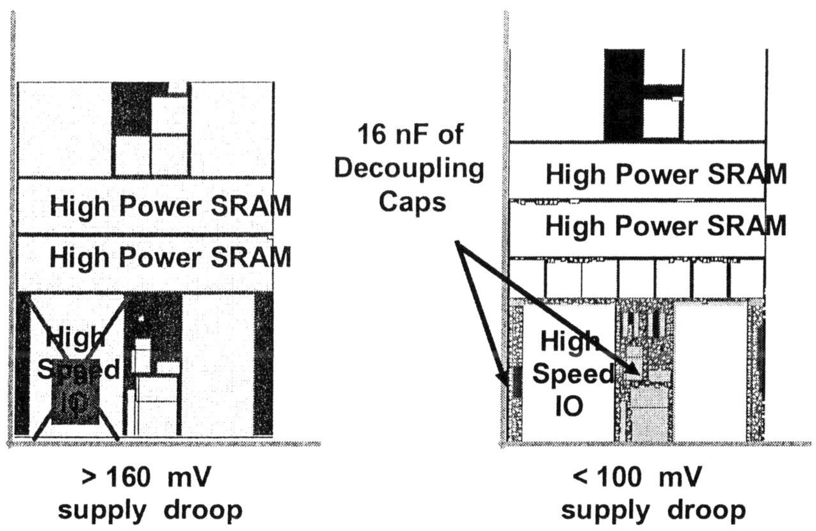

Figure 11-14. Power supply noise example

2.3.3   Substrate noise

In addition to line-to-line and power supply coupling, substrate noise injection remains an important signal integrity concern. Substrate noise results when high speed device switching causes overshoot current to be injected into the substrate of a switching SoC. Since all components of an SoC share the same substrate, the substrate becomes a signal path for unwanted noise. This is particularly problematic in mixed signal designs where sensitive analog circuits must function in close proximity to fast switching digital logic. That being said, an increasing number of future SoCs will be mixed signal chips.

Mixed signal designers have long relied on separate power supplies and substrate isolation structures such as *guard rings* to reduce the impact of substrate noise. Guard rings work by shunting substrate currents such as those introduced by noisy digital switching and prevent them from interfering with sensitive analog circuitry.

3.   PARAMETRIC VARIABILITY

Another factor of increasing concern to SoC developers is process variability. Despite advances in every phase of processing, there remain uncontrollable variations in the dimensions and material properties of the finished product. The most studied variation is *across chip linewidth variation (ACLV)* which creates small perturbations in the critical gate length of devices. ACLV has many causes including uneven etching due to local shape density variations as well as lithographic distortions caused by optical interactions between shape regions. Another source of increasing variation is *micro-implant variability* which is caused by small changes in ion implantation dosage across the wafer. These small variations can cause identically designed devices to have significantly different electrical characteristics. These variations may be observable both when measuring the same device on different chips (*inter-die*) or between identical devices on the same chip (*intra-die).*

Traditionally, it has been sufficient to add a small safety margin or *guard band* around performance specifications to ensure that the large majority of chips will function. As process dimensions and supply voltages shrink, the relative impact of these variations is increasing, as shown in Figure 11-15 which shows variation trends in geometric factors such as effective device length (L_{eff}), device width (W), oxide thickness (T_{ox}), operating temperature (T), supply voltage (V_{dd}), etc.

Both inter-die and intra-die variations cause the actual performance of a given chip to deviate from the intended performance. For example, Figure 11-16 [21] illustrates inter-die variations as measured using identical ring oscillators placed on each die on a 200 mm silicon wafer. Areas of identical shading have identical frequency measurements. The total range of variation is 30%.

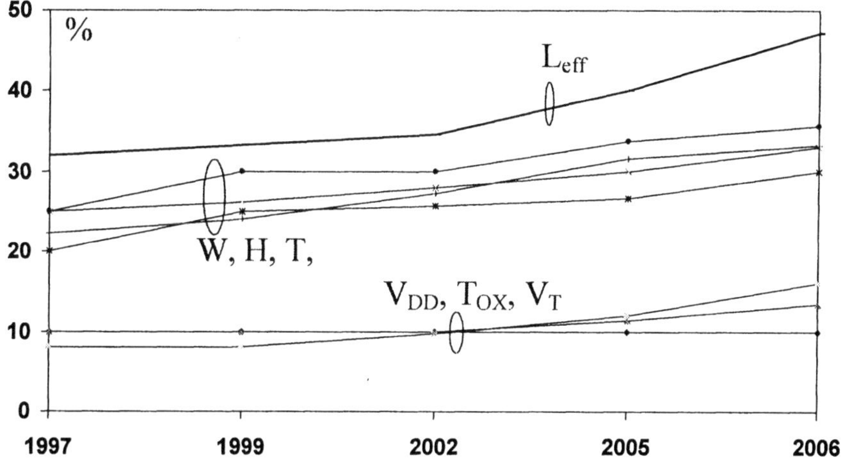

Figure 11-15. Variability trends

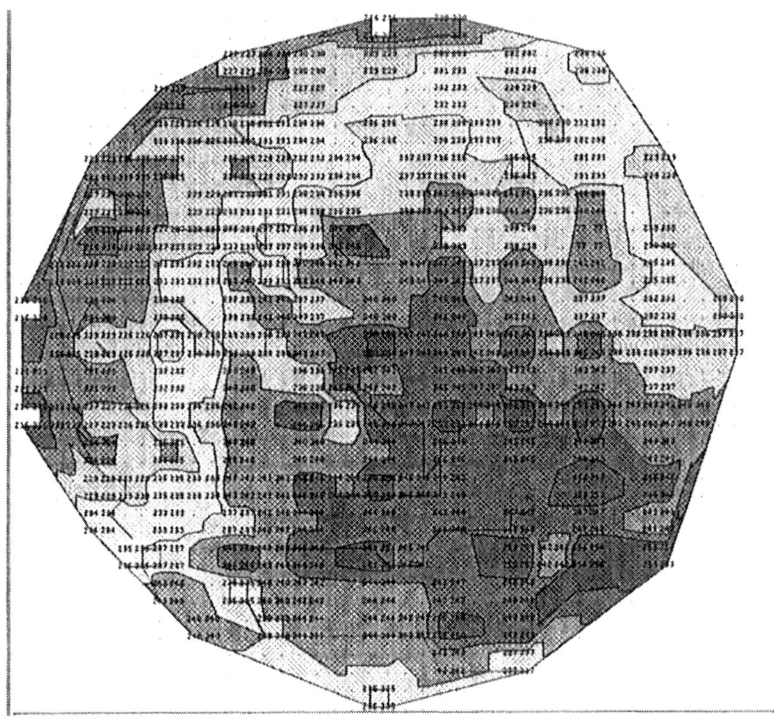

Figure 11-16. Wafer-map of inter-die parametric variations on ring oscillator frequency

The impact of these increased variations on SoC design is profound. First, an increasing amount of over-design will have to be built into every design to ensure that it will work over all ranges of parameter variation. This implies that an increasing amount of performance is being sacrificed for each subsequent technology generation.

In addition to sacrificing performance, increasing statistical variation is complicating the analysis of complex SoCs. To be certain that a chip will function over the complete range of parametric variations can require multiple performance analyses. For example, it may be necessary to analyze chip timing by simultaneously assuming both the fastest devices due to short L_{eff}, with the slowest wires due to thinning of the interconnect layer; and analysis must also be performed at the opposite extreme. In general, every statistically independent parameter must be analyzed at the worst case and best case end points of its parametric range. This effectively multiplies the number of timing analyses by a factor of two for each modeled variation. In cases where the effect of the parameter on timing is not monotonic, even this method can be insufficient to ensure timing correctness.

There are several techniques that can be used to minimize variation in SoC design. Because the relative impact of parametric variability is worse with smaller features, designers can improve device to device matching by avoiding minimum-sized structures wherever possible. Since many sources of variability have a directional component, giving all critical components the same alignment can dramatically improve matching [22].

One of the best weapons against ACLV is uniform shape density. By ensuring that all regions of the chip contain the same density of polysilicon shapes, etch-induced dimensional variation can be reduced. In addition, shape regularity improves the matching of critical shapes by minimizing the difference in their optical interactions. The rule for reduced variability is always: the more regular the pattern of shapes, the better the dimensional control.

The next important step in coping with increasing parametric variability will be a move towards more statistically-aware design tools. Several promising prototypes of statistical timing tools have already been demonstrated [23]. Such tools will allow more accurate predictions of hardware performance distributions, which will ultimately lower the amount of over-design required, improve power performance and increase manufacturing yield.

4. RELIABILITY

A final design consideration for SoCs is long term functional reliability. Reliability concerns are related to processes that may allow a chip to function correctly immediately after manufacturing but may cause the chip to malfunction at some later point during its working life. In the best case, this type of unexpected chip failure may present a costly inconvenience to the customer. In other, more mission-critical applications such as automotive, avionics or security, the result of a malfunction can be life-threatening. There are many reliability factors that can affect the long term operation of a chip. Most fit into one of three categories: those that degrade the characteristics or function of a device (*device wear out),* those that degrade the characteristics or function of interconnect (*interconnect wear out*) and those that cause transient disruption of function.

4.1 Device wear out

There are several device wear out functions that can cause long term change in the function of a device on an SoC. Two common concerns are Hot Carrier Injection (*HCI*) and negative bias threshold instability (*NBTI*). HCI occurs when electrons in the channel of a transistor are accelerated by the high electric field found near the drain of devices which are biased near pinch-off. These highly energetic electrons ionize bonds in the channel gate interface where they create electron or hole traps. Over time these traps lead to charge build-up in the gate which effectively causes a threshold voltage shift for the device. HCI is accelerated for devices which have high applied gate voltages, high switching activity and/or drive large loads. NBTI is similar in effect, but does not require high electric fields. It affects both switching and non-switching p-channel devices. NBTI is accelerated by high operating temperatures and can be induced during burn-in test [24]. Another wear out mechanism is *time dependent dielectric breakdown (TDDB)*. TDDB is a gradual breakdown of the gate dielectric caused by high electric fields over extended periods.

The effects of NBTI and TDDB are increasing over time due to the use of thinner gate oxides required by scaling. As these effects increase, the impact on SoC design is also increasing. Shifting the threshold of devices has a direct effect on the delay through a device. Delay can either increase or decrease with time depending on the nature of the injected charge and the type of device. In time the performance shift may be large enough to cause the chip to malfunction.

To ensure that device wear out does not cause functional problems, it is important to avoid over stressing devices by extended high voltage and high temperature. It is also becoming increasingly essential to predict the worst case threshold shift which might be induced over the working life of the SoC and include this shift as a design margin. In this way the design can be demonstrated to work with both start- and end-of-life device characteristics. The over design required to ensure this, of course, comes at a cost in performance.

4.2   Electromigration

Like device wear out, there are mechanisms that can cause a shift in the characteristics of chip interconnect with time. The principle mechanism is *electro-migration (EM)*. EM occurs when ballistic collisions with flowing carriers causes the metal atoms in interconnect to migrate away from their original position.   For this reason EM is much more an issue in wires carrying uni-directional (DC) rather than bidirectional (AC) current. Electromigration is accelerated both by increasing current density and by increasing temperature as shown in Figure 11-17.

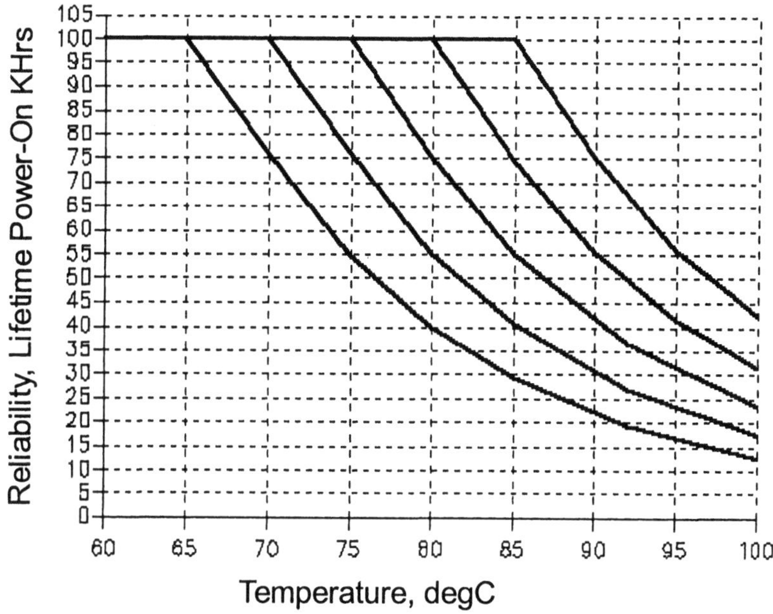

Figure11-17. Temperature induced interconnect wearout

Several trends are causing the effect of electromigration on SoCs to worsen with time. Decreased wire cross sections required by scaling and non-scaling wiring linear thickness have increased wire current density. Additionally, the new low permittivity dielectric materials introduced to help performance have inferior thermal characteristics than the materials they replaced. The result is an increase in wire *self-heating* which further accelerates EM.

This movement of metal caused by EM can cause thinning of wires to the point that their resistance increases, metal opens or increases in contact resistance as is shown in Figure 11-18.

To minimize the impact of device electromigration, SoC designers must take steps to manage the current density and operating temperature of interconnect. These steps should include using wider wires on signals that must carry higher switching or DC currents. This is particularly important in analog design where DC currents are frequently quite high. In addition any steps in power optimization or packaging that can reduce chip operating temperatures can also help reduce EM.

Electromigration wire fail   **Electromigration contact fail**

Figure 11-18. Electromigration fails

4.3 Transient reliability problems

In addition to the gradual wear out processes just described there are several processes which can suddenly affect the function of an SoC. The first is *electrostatic discharge* (*ESD*) and the second is *soft errors (SE)*.

4.3.1 ESD errors

ESD is a transient condition in which unacceptably high voltage is inadvertently presented to an input of the chip during handling, installation or improper grounding of equipment and cables. When an ESD event occurs, the gate of any devices connected to the transient high voltage is destroyed due to the high field induced in its gate oxide. In some cases the wiring that connects the pin to the device may also be destroyed due to the high induced transient currents. Because of the permanent damage done, ESD is generally a non-recoverable error.

SoC designers can protect their chips from ESD by using properly sized ESD protection networks on all input signals which enter a chip. These structures generally consist of clamping diodes and a small series resistor which allows the transient voltage to be safely dissipated into the substrate.

4.3.2 Soft errors

Soft errors are transient errors which are induced by high energy charged particles which either originate from outer space or from nearby radioactive materials. The carriers induced by the charged particle as it travels through the silicon substrate of the SoC can disrupt the logic state of sensitive storage elements. For example, *dynamic read-only memories* (DRAMs) store their information as a small charge on a small capacitor. If a fast alpha particle hit occurs near enough to the DRAM storage node, the small charge stored there could be reduced. If this effect is great enough the '1' stored in the DRAM cell might be accidentally changed to a '0', effectively destroying the information stored there. Note that these errors are called *soft errors* because while they do disrupt data, they do not permanently damage the SoC.

As scaling decreases DRAM cell sizes, the critical *charge* (Q_{crit}) needed to preserve the intended state is decreasing, making DRAM ever more susceptible to this type of soft error. Similarly, the very small devices used in *static read-only memory (SRAM)* are also susceptible to SE upset. In fact, the SE rates for SRAM are now considerably worse than those for DRAM in the same technology. Some projections are predicting that SoC logic and latch structures might also soon be susceptible to particle-induced soft errors. Such latch errors have already been observed in advanced technology for microprocessors.

There are several things that SoC designers can do to minimize the effect of SE on their most critical designs. The most basic precaution is the addition of *error correction code* logic (*ECC*) into critical storage elements such as DRAMs and SRAMS. ECC logic uses additional storage bits to encode parity information that can be used to detect and correct single or multi-bit failures due to soft error upset. This added security comes, of course, at a cost of extra array size and power requirements. When predictions regarding SER errors in SoC logic and latching come true, more extreme measures will become necessary. Some researchers have proposed dual rail or other redundant logic schemes to provide circuit level parity checking on sensitive logic. For the most critical applications it is possible to provide redundant functions at a system level and compare the results in a simple voting scheme.

5.   MANUFACTURABILITY

In addition to facing increased challenges on the design side, chip producers are facing a significant increase in the complexity of semiconductor manufacturing. Two manufacturing issues of particular interest to SoC designers are manufacturing yield and new lithography challenges.

5.1.1   Yield issues

For all its sophistication, not everything relating to semiconductor manufacturing is an exact science. Parametric variations or random defects can be introduced at almost any step of manufacturing which can cause a chip not to function as intended, The more chips affected by manufacturing induced errors, the fewer good ships are yielded per wafer. The main factor that contributes to SoC yield loss is *foreign material defects*. These defects result when small bits of material accidentally fall on the mask reticule or the photo emulsion and interfere with the proper creation of a chip structure. These defects may take the form of unintended shorts between adjacent structures, unexpected holes in insulating materials such as device gates or, unintended opens in a conductor. In some case the defects cause partial opens which lead to high resistance connections. The actual yield of a chip is given by Equation 4.

$$Y = (1 + \frac{Ad}{\alpha})^{-\alpha} \tag{4}$$

Where Y is the yield, or the probability of a good chip, A is the area of the chip, d is the defect density, and α is an empirical defect clustering parameter. The defect density, d, is determined by two main factors: the concentration and size distribution of the foreign material in the processing environment and the spacing between structures on the chip. The regions of the chip where the spacing is less than or equal to the expected size of the largest foreign material are called the *critical area*. Designs with large amounts of critical area generally have lower yield.

To the great credit of process engineers, foreign material size distributions are being reduced at the same rate as object spacing is being reduced by scaling. Figure 11-19 illustrates the defect density trends at IBM over a long period. Notice how that after a short yield learning period, the defect densities in each new generation return to the scaling trend line.

Unfortunately, managing defect density is not enough to maintain good yields. Each process technology is getting more complex with scaling, with the addition of extra routing layers, and the use of new materials. In addition, advances in design tools and methodologies are increasing circuit density. The net result is that chip critical area is slowly increasing with each process generation. With chip sizes roughly constant, this also implies that yields are also getting worse with each new process.

The main impact of decreasing chip yield is an increase in manufacturing cost. The fewer good chips produced per wafer, the more expensive those good chips will be. Another impact is on the nature of testing. An increasing number of defect types create resistive shifts rather than complete opens or shorts. These marginal defects can cause subtle changes in timing and may not all be caught through standard reduced speed *stuck-at* testing. The best way to reliably screen out these types of defects is to introduce some sort of AC or *at speed* testing to catch defect induced *delay faults*.

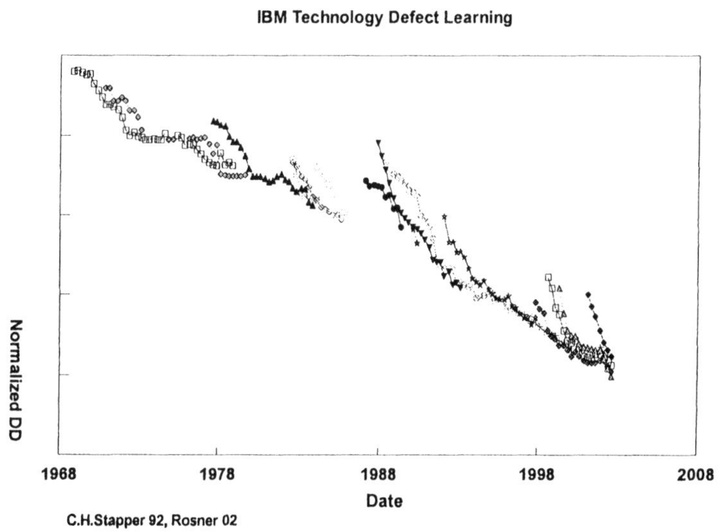

Figure 11-19. IBM Normalized Defect Density vs. Time

There are several techniques that an SoC or IP developer can use to maximize yield for his or her design. First, minimizing chip area increases yield directly by lowering the likelihood of a defect falling on a given die. In addition, the smaller the chip, the larger the number that can be put on a given wafer. The two factors combine to greatly lower production costs for smaller chips. Second, defect density can be reduced by selectively increasing inter-device and inter-wire spacing to be larger than the expected size of foreign material in the process. This reduction of critical area can be accomplished by using relaxed spacing ground-rules in most portions of the chip while resorting to minimum spacing rules in only the most space constrained regions. For example, new routing techniques have been developed such as those in IBM and U. of Bonn's Xrouter, which preferentially avoids putting wires in adjacent tracks where possible. Other techniques such as IBM's WireBender [25] improve yield by inserting additional inter-wire space in completely wired layouts.

Finally, the most powerful technique for improving yield is redundancy. By creating designs that can tolerate a small number of defects and still function, yield can be dramatically improved. Redundancy can come in many forms: at its lowest level, the use of redundant device contacts and interlayer vias; at an intermediate level, redundancy in DRAM or SRAMS; and at a system level, by architecting in system level redundancy, an SoC can still be functional even when an entire IP block is defective. In one example IBM built a large SoC with 16 identical processor clusters. Through re-architecting, allowing a single bad cluster in the middle block of eight processors increased yield by 50%, and allowing any of the 16 clusters to be faulty doubled the manufacturing yield.

5.1.2   Lithography issues

A second source of manufacturing concerns for an SoC designer is lithography evolution. The principle driver of silicon scaling has been advances in the lithographic processes used to transfer design geometries to a manufacturing wafer. In an ideal world, this transfer would be ideal; i.e., design shapes would be transferred *as-is* to the photolithographic mask, the mask would transfer its image undistorted to the photoresist on the wafer and the wafer processing would faithfully reproduce the image on the mask. The reality is unfortunately quite different from this ideal. Every step in the transfer of design information to finished patterned wafer structures is subject to distortion and disruption. There is a fundamental limit on the minimum feature size that can be imaged using a particular lithographic setup. This limit is captured in the *Rayleigh* equation below.

$$R = k \frac{\lambda}{NA} \qquad\qquad (5)$$

Where R is the *Rayleigh* number or the smallest dimension that can be imaged, k is the so-called *k-factor,* λ is the illumination wavelength, and NA is the numerical aperture or lens size of the stepper.

In the past, the ability to image ever smaller shapes has been driven by reducing λ that is using shorter wavelength illumination in the stepper tools used to transfer images from mask to wafer. Unfortunately, each subsequent jump in λ is comes at great increase in cost and technical complexity. With the industry struggling to bring up tools at $\lambda = 157$ nm, the likelihood of continuing scaling through further decreases in λ look bleak. Figure 11-20 shows the trends in λ scaling and its impact on k.

The impact of a stall in the easy reductions in λ has a potentially profound impact on SoC design. The further that dimensions shrink into this *sub-resolution* regime, the more distorted shapes due to *diffraction effects* during imaging. This trend is shown in Figure 11-21. Unchecked, this trend would ultimately limit the pace of SoC scaling.

The industry is increasingly looking at the *k-factor* as the best hope to enable continued scaling. The *k-factor* is determined by a number of factors including mask materials, mask thickness and the shape and relative position of the mask shapes themselves. By manipulation the mask image such that diffraction effects selectively cancel it is possible to lower the *k-factor* from its theoretical limit of approximately 0.5 to 0.25 or below.

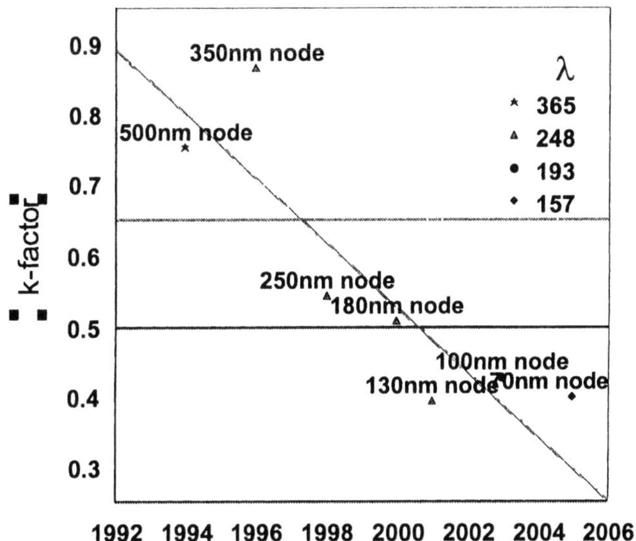

Figure 11-20. Trends of lithography *k-factor*

There are three major *resolution enhancement techniques (RET)* being introduced to achieve this improved imaging for SoCs: *optical proximity correction (OPC), alternating phase shift masks (alt-PSM),* and the use of *sub-resolution assist features (SRAFs).*

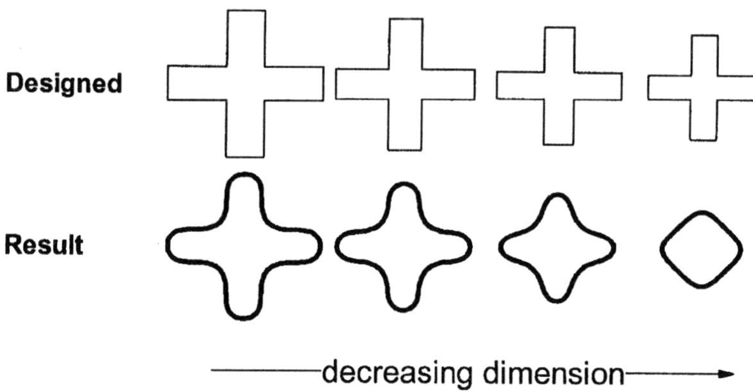

Figure 11-21. Sub-resolution diffraction distortion

In OPC designed mask shapes are modified in such a way by a preprocessing step that anticipates and partially compensates for the distortion of the final imaging. In this way the actual imaged shapes more closely resemble the design shapes [26]. Figure 11-22 illustrates this process. The image on the left shows both the mask shapes and the desired image as the solid white rectilinear shapes. The grey filled shape shows a simulated image of the distorted shape that would be produced through manufacturing. Notice how distorted this resulting image is compared to the desired mask image. The right image, in contrast, shows the result of OPC processing. Again, the solid white rectilinear shapes show the image on the mask. Notice the notches and protrusions that have been added to the mask shape by the OPC pre-processing. Again, the grey filled shapes show the shapes that would result from manufacture. Notice that the imaged shapes much more closely reproduce the intended geometry.

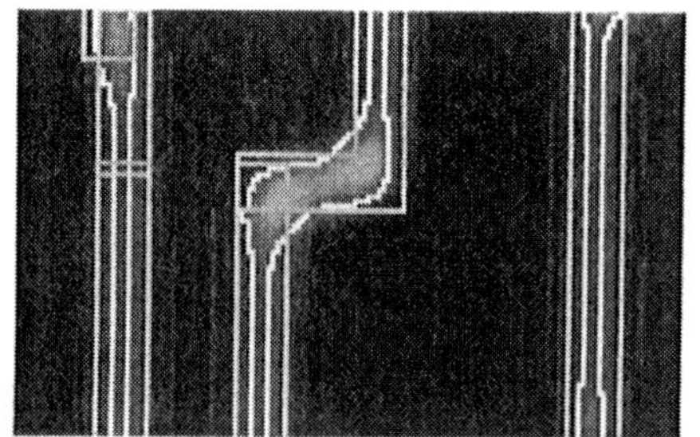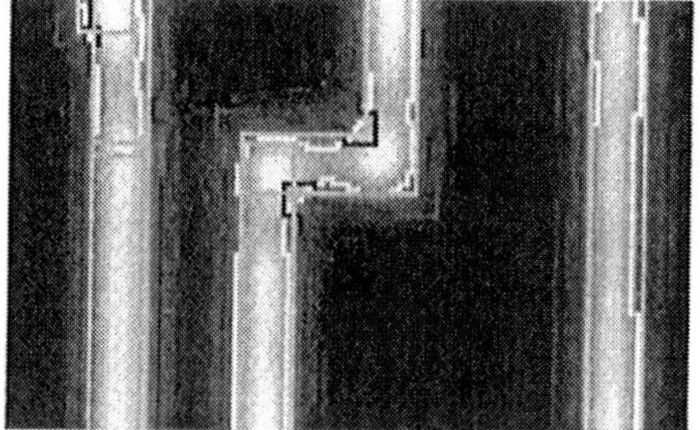

Figure 11-22. Lithographic fidelity without and with OPC

Another technique used to improve sub-resolution imaging is *alternating phase shift mask (alt-PSM)* generation. Like OPC, alt-PSM attempts to pre-process mask data to improve the fidelity of printing. In this case, it is the mask thickness rather than the design shapes that are modified. By selectively thickening and thinning the transmissive portions of the mask, the phase of the illuminating light can be manipulated to negate the effects of diffraction distortion. [26].

In addition to adding complexity to the mask making process, alt-PSM has a significant impact on the design process. To function, alt-PSM requires that all portions of a mask be *colored* into regions of alternating phase. This requires that no two regions of the same phase are adjacent at minimum spacing. This can limit the type of geometry and the spacing used during layout. For these reasons the introduction of alt-PSM is being delayed as long as possible. Industry consensus is that alt-PSM will be needed for critical polysilicon dimension control in the 90nm node and beyond.

A final technique is the addition of sub-resolution assist features or SRAFs. Like alt-PSM, SRAFs attempt to manipulate interference effects. In this case, however, interference is managed through the addition of sub-resolution mask shapes or *correctors*. These correctors are shapes that appear on the mask, but are too small to be imaged on the wafer. They only act to selectively diffract the mask illumination to improve imaging. [26].

Like alt-PSM, the use of SRAFs can impact design. Layout must be structured in such a way that allows sufficient space for correctors to be included. Again this can cause geometry restrictions which can impact density and ultimately could affect designer productivity.

All of these RET solutions can have a big impact on SoC design. In addition to the density and productivity issues already mentioned, these create another problem which must be simultaneously solved with performance, power and signal integrity closure. The only way to approach this in the future will be the creation of manufacturing aware design methodologies.

6.   AFFORDABILITY

While cost has always been an issue in the creation of SoCs, process complexity has reached a point that cost of design may be a bigger limiter than physics in SoC performance and function. There are two types of cost that are affecting the business of SoCs: the *non-recurring engineering (NRE)* costs associated with design and the manufacturing costs themselves.

6.1   NRE Costs

NRE costs are increasing for a number of reasons. Increases in both chip and process complexity are making design of a modern SoC ever more expensive endeavor. New designs require larger, more skilled design teams equipped with more sophisticated and expensive design tools. The increased difficulty of design closure requires longer design time, and often more costly design iterations. Verification is also becoming much more difficult as system complexity increases. Many of these costs associated with the designer productivity gap are dealt with elsewhere in this volume. Technology issues also play a big role in increasing NRE. With mask costs increasing to more than a million dollars per design pass, and the average number of design passes required per SoC hovering between 2 and 3, designers may soon find it difficult to justify custom SoCs for all but the highest volume applications.

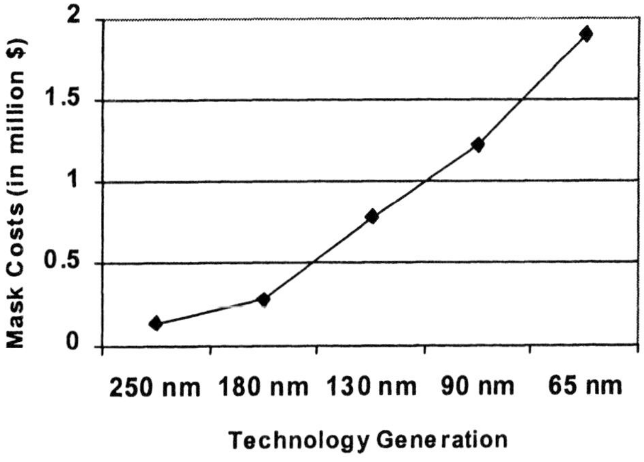

Figure 11-23 Mask cost trends

High NRE costs are causing SOC designers to consider alternatives for their system design needs. For designs where density and performance are less critical *field programmable gate arrays (FPGAs)* are becoming an increasingly attractive alternative to ASIC based SoCs. FPGAs offer much lower NRE costs than traditional SoCs because they require no mask build and have much simpler, albeit much more constrained design methodologies. Despite the much higher unit costs of FPGAs, the total cost of production for low volume applications can be lower due to these NRE savings.

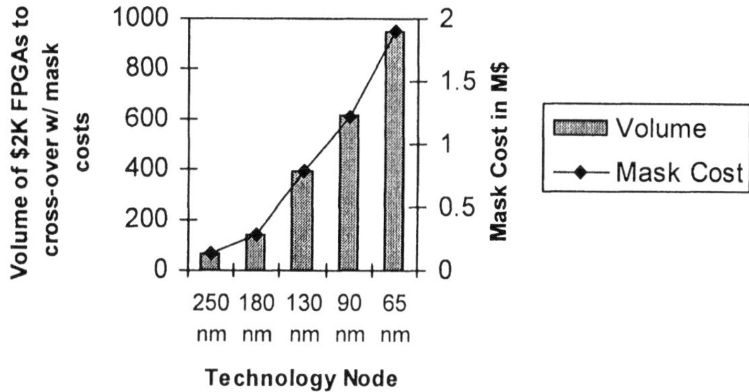

Figure 11-24. FPGA volume cross-over points with ASIC mask costs

There are significant issues associated with using FPGAs for SoC design. There is up to a 10X cost in power for using FPGAs when compared with ASICs. FPGAs are also considerably slower than high-end AISC-based SoCs. Figure 11-25 illustrates the power-performance values for both technologies.

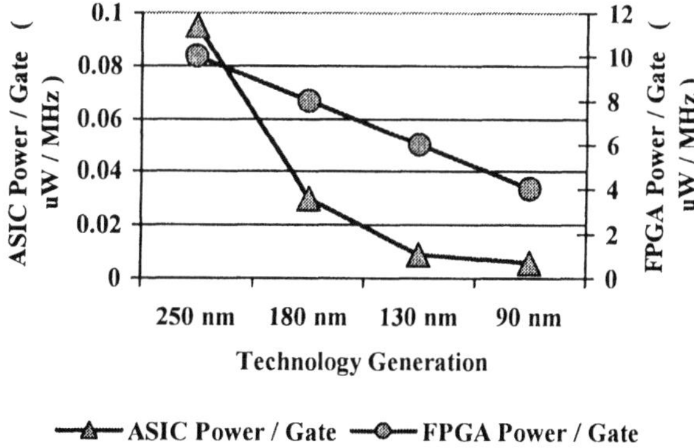

Figure 11-25. ASIC vs. FPGA Performance Comparisons

In response to these issues, the industry is developing a number of hybrid solutions that couple the performance and power advantages of ASIC design style with the low NRE and re-spin costs of FPGAs. These *embedded FPGA* offerings take the form of a standard *platform* which couple ASIC style SoC IP such as processors, memory and high speed serial IO with programmable

FPGA fabric for customer logic. With this mix it is now possible to use the programmable circuitry to enable a single physical chip design to satisfy several different applications. This has the potential to eliminate multiple designs and in some cases, avoid costly re-spins. In the case where a customer requires several similar ASICs for a family of products, FPGA circuitry can be added to the base ASIC logic and be configured as needed to satisfy the multiple applications. Similarly, logic updates required to correct bugs discovered late in the verification process, or to accommodate changing market needs can be handled with appropriately placed FPGA cores [1].

Another emerging alternative to full SoC design are hybrid SoC platform solutions with embedded gate array. These SoCs combine a predefined set of hard IP with user customizable gate array logic. The customer then only pays NRE charges for the metal masks used to personalize the gate arrays [27].

6.2   Technology Integration Costs

A final manufacturing cost concern for SoCs relates to increasing cost due to technology integration. One of the main drivers of SoC has been the desire to mix multiple technologies on a single silicon die. The trend has been to integrate what might have been several chips in different technologies onto a single chip. Doing this can reduce the cost of design by lowering packaging cost, it can improve performance by avoiding the delay overhead of chip-to-chip interconnect and it can reduce the board space required in a space constrained system such as a cell phone. In spite of the obvious advantages, technology integration brings great challenges of process compatibility, electrical compatibility and cost.

Process compatibility issues relate to integrating features from two or more separate technologies into a single process. As scaling drives up process complexity and drives down supply voltages process integration requires significant trade-offs in technology capabilities. For example, integrating DRAM with fast ASIC logic involves compromises in substrate doping. The optimal substrate doping needed for fast logic is not well optimized for DRAM process. As a result embedded DRAM generally has lower performance and needs more frequent refresh than stand-alone DRAM [2]. Similarly, process optimizations and supply voltage limits for high performance logic are often at odds with those which would optimally be used for high precision analog design [28].

Technology integration also raises issues of electrical compatibility. Signal integrity becomes a significant concern when placing sensitive analog or DRAM elements onto the same chip as fast logic. Mixed IP integration also raises special challenges for system test. For example a mixed signal SoC will require a single test flow that incorporates scan based logic test with at-speed analog functional test. This can drive either higher test equipment cost or throughput issues related to the use of multiple testers.

One increasingly attractive alternative to SoC is system on a package (SoP) (or System-in-Package (SiP)). Through advances in packaging technology, the traditional electrical overhead of chip to chip interconnect has been reduced. It is now possible to integrate multiple dies together using interconnect that approaches the density and delay characteristics of on chip interconnect [29]. Advances in chip stack technology [30] have made it possible to vertically stack multiple dies using specially modified flip chip or wire-bond techniques as shown below in Figure 11-26.

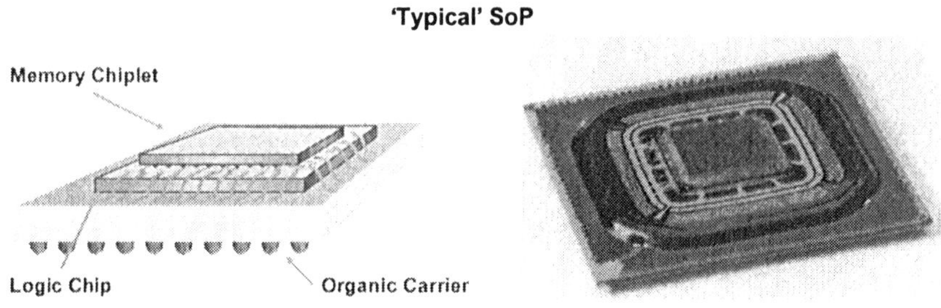

Figure 11-26. SoP Examples

Both these package-level and chip stack alternatives allow individual system components to be built and separately yielded in the best available technology. The cost savings associated with being able to separately yield theses smaller *chiplets* can in some cases offset the increase in complexity of multi-chip packages. Consider the case in Figure 11-27 where a flip chip implemented SoP combination of a 50 mm^2 logic chip and a 50 mm^2 DRAM was sized to be roughly half the cost of a functionally identical system requiring a 95 mm^2 mixed logic DRAM SoC.

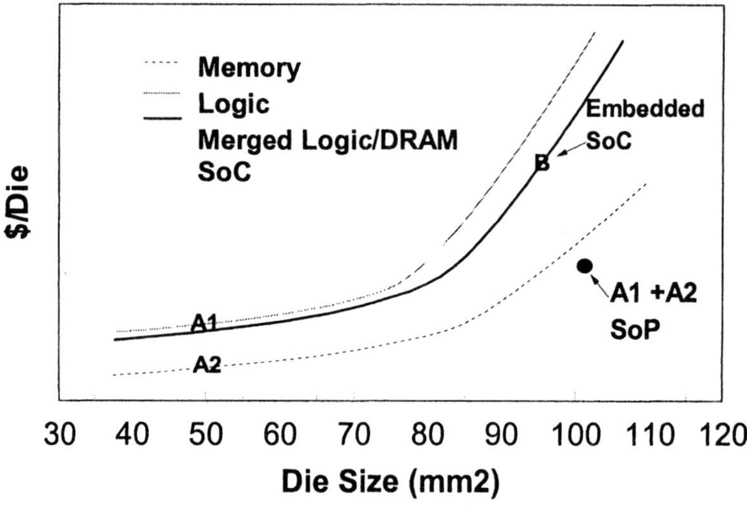

Figure 11-27. SoP vs. SoC cost comparison

Because the system design considerations of SoP are essentially the same as those for SoC, much of the infrastructure and IP investment made for SoC can be reused.

7.   SUMMARY

This chapter has provided a high-level overview of the technology related challenges facing SoC designers today. These challenges have been shown to impact the design, manufacturability and ultimately the overall affordability of current and future SoCs. The list of challenges is certainly long and difficult. While it may not be possible to say with certainty how and when each of the problems presented will be solved, it is certain that the innovation that has fueled the growth of SoCs to this point will continue to find solutions to even these thorny problems.

ACKNOWLEDGEMENTS

The author would like to acknowledge the help of a great group of people including: Kerry Bernstein, Ron Rose, Tom Bednar, Paul Zuchowski, David Hathaway, Ed Nowak, Gus Tellez, Joe Kozhaya, Ivan Wemple, Ray Rosner, Dan Maynard, Mark Lavin, Chandu Visweswariah, Sani Nasif, Brian Zoric, Dale Hoffman, David Lackey, Larry Pillegi, Rob

Rutenbar, Ruchir Puri, Leon Stok, Kathy McGroddy-Getz, Grant Martin, and Steve Teig. All contributed materials and and/or participated in helpful discussion related to this section. My special thanks to Diane Mariano, for contributions in form, content and emotional support.

REFERENCES

1. Paul S. Zuchowski et. al., "A Hybrid ASIC and FPGA Architecture", *ICCAD 2002*, November, 2002, pp. 187-194.
2. S. Crowder, R. Hannon, H. Ho, D. Sinitsky, S. Wu, K. Winstel, B. Khan, S.R. Stiffler, and S.S. Iyer, "Integration of Trench DRAM into a High Performance 0.18um Logic Technology with Copper BEOL", *International Electron Devices Meeting, 1998*, pp. 1017-20.
3. http://www-3.ibm.com/chips/products/asics/products/low-k.html/.
4. K. Chuang and R. Puri, "SOI Digital CMOS VLSI - A Design Perspective", *36th. Design Automation Conference*, June, 1999.
5. URL: http://news.com.com/2100-1001-268083.html/.
6. J. Kedzierski, et. al. "Metal-Gate FinFET and Fully-Depleted SOI Devices Using Total Gate Silicidation", *IEDM 2002*.
7. Ron Ho, Ken Mai, Herna Kapadia and Mark Horowitz, "Interconnect scaling implications for CAD", *ICCAD 1999*, pp. 425-429.
8. URL: http://www.chips.ibm.com/techlib: 128bitPIbBus.pdf/.
9. Girish Varatkar and Radu Marculescu, "Traffic Analysis for On-Chip Network Design of Multimedia Applications", *39th Design Automation Conference*, June, 2002.
10. M. Igarashi, T. Mitsuhashi, A. Lee, et. al., "A Diagonal-Interconnect Architecture and its Application to RISC Core Design", *ISSCC 2002*.
11. URL: http://www.xinitiative/.org/wt/home_flash.php/.
12. E.J. Nowack, "Maintaining the Benefits of CMOS Scaling when Scaling Bogs Down", *IBM Journal of Research and Development*, Number 2/3, March/May 2002.
13. URL: http://public.itrs.net/Files/2002Update/2002Update.pdf/.
14. International Technology Roadmap for Semiconductors (ITRS), 2002 edition, *System Drivers* Chapter.
15. Wolfgang Nebel, *Power Estimation at the Logic Level*, Kluwer Academic Publishers, Dordrecht, Netherlands.
16. "IBM lowers power 10X on PowerPC 405", *EETimes*, October 15, 2001. URL: http://www.eedesign.com/isd/OEG20011011S0076/.
17. B.R. Stanisic, N.K. Verghese, R.A. Rutenbar, L.R. Carley, and D.J. Allstot, "Addressing Substrate Coupling in Mixed-Mode ICs: Simulation and Power Distribution Synthesis", *IEEE Journal of Solid-State Circuits*, volume 29, March, 1994, pp. 226-237.
18. John Darringer et. al., "EDA in IBM: Past, Present and Future", *IEEE Transactions on CAD*, Volume 19, number 12, December, 2000.
19. D.E. Lackey, P.S. Zuchowski, T.R. Bednar, D.W. Stout, S.W. Gould, and J.M. Cohn, "Managing Power and Performance for System-on-Chip Designs using Voltage Islands", *International Conference on Computer-Aided Design (ICCAD)*, San Jose, California, November, 2002.
20. S.R. Nassif and J.N. Kozhaya, "Fast Power Grid Simulation", *37th. Design Automation Conference*, Los Angeles, June 2000, pp. 156-161.

21. K. Bernstein, "Design, process and environmental contributors to CMOS Delay variation", *SSCTC Workshop*, 2003.
22. S. Nassif, "Modeling and Forecasting of Manufacturing Variations", *Asia-Pacific DAC*, 2001.
23. C. Visvesvariah, "Death, Taxes and Falling Chips", *Proceedings of TAU 2002*.
24. Y. Mitani, et. al., "NBTI Mechanism in Ultra-thin Gate Dielectrics - Nitrogen-Oriented Mechanism in SION", *IEDM 2002*.
25. US patents: US 6,305,004: Method for improving wiring related yield and capacitance properties of integrated circuits by maze-routing, Gustavo Tellez et. al.; US 6,189,132: Design rule correction system and method, F. Heng et. al.
26. M. Lavin and L. Leibman, "CAD Computation for Manufacturability: Can We Save VLSI Technology from Itself?", *ICCAD 2002*.
27. URL: http://www.gigascale.org/pubs/reports/reports/GSRC1Q02Web.htm/.
28. M. Hatamian, et. al. "Design Considerations for Gigabit Ethernet 1000Base-T Twisted Pair Transceivers", *Custom Integrated Circuits Conference (CICC) 1998*.
29. H.B. Pogge, "The Next Chip Challenge: Effective Methods for Viable Mixed Technology SoCs", *Design Automation Conference*, 2002.
30. J. Dufresne, S. Ouimet and T.R. Homa, "Hybrid Assembly Technology for Flip-Chip-on-Chip (FCOC) Using PBGA Laminate Assembly", *ECTC'00: Electronic Components and Technology Conference*, 2000.

Index

Accellera 188, 195
Alba 4, 7, 8, 9, 11
Altera ix, 16, 26, 27, 29, 37, 159, 164, 185
AMBA 187-189, 195, 196, 198, 199, 202- 206, 208, 209, 220- 222, 226
AMBA High-Performance Bus 108, 162, 199, 208, 209
Analogue/Mixed-Signal (AMS) iii, 5, 6, 21, 22, 33-36, 38, 45, 59, 67, 173, 233, 241
Application Software 15, 28, 55, 138, 151, 187- 190, 198, 207, 208, 210- 213, 216, 223, 225
ARC 10, 16
ARM ix, 10, 11, 15, 23, 25, 26, 30, 40, 116, 118, 143, 162, 187, 188, 190, 193, 195-198, 202, 203, 204, 206, 212, 213, 219, 220, 222- 227
Avalon 159, 162, 174, 185

Berkeley Wireless Research Center (BWRC) x, 16, 26, 30, 229, 252
BOPS 10
Bricaud 4, 7, 19, 66
Bus Protocol 31, 33, 57, 74, 121, 198, 201, 208
BWRC Emulation Engine (BEE) 231, 236- 245, 249

C++ 28, 57, 61, 96, 217
Canada iii, 7
Certification 41, 114, 223
Co-Design 24, 26, 40, 44, 45, 79, 159, 226, 227, 253
Consumer Electronics 67, 71, 84
CoreConnect 120- 123, 128, 133, 147, 150, 151, 154

Decomposition 57, 58
Design Convergence 56
Design Methodology 1, 21- 25, 34, 35, 38, 41, 157, 159, 258
Development Working Group iii, 5, 12, 19, 58, 59
Digital Video Platform 67

Electronic Design Automation (EDA) 3, 4, 5, 6, 17, 44, 108, 113, 138, 196, 205, 219, 220, 229, 295
Embedded Software iii, vii, 6, 8, 21, 22, 24, 25, 27, 28, 29, 32, 38, 136, 137, 150, 188, 191, 197, 199, 201, 207, 210, 226
European Chips and System design Initiative 5

Field of Experience 43
Field of Use 43
Field-Programmable Gate Array (FPGA) 19, 26, 27, 29, 134, 135, 141, 142, 144, 147-151, 154, 156, 159, 160, 163, 168, 178, 180, 184, 207, 211, 212, 223, 229- 231, 238, 240-242, 244, 249, 250, 252, 253, 291, 295
Flip Chip 39, 293
Foundry ix, 7, 47, 48, 53, 62- 65
French Revolution 9, 12, 15
Front-End Acceptance 22, 25, 35, 42
Fujitsu 5, 61

Hardware Emulation 229, 237, 239, 245, 252
Hardware-Dependent Software 188, 189, 190, 198, 206, 207, 212, 215, 222, 223

IBM i, ix, x, 11, 23, 25, 30, 34, 36, 40, 119, 120, 124, 126, 131, 133, 135, 137, 138, 146, 150, 255-258, 267, 271, 284-286, 295
IC Design 3, 8, 9, 12, 21, 34, 38, 39, 40, 53, 56, 58, 232
Infrastructure 8, 10, 11, 21, 22, 37, 38, 76, 95, 100, 115, 139, 151, 152, 184, 190, 198, 205, 209, 211, 212, 219, 237, 241, 294
Institute for System-Level Integration 8, 11
Integrated Device Manufacturer (IDM) ix, 62
Intellectual Property (IP) ii, iii, vii, viii, ix, 1, 3- 15, 18, 19, 22, 23, 30, 32, 35-38, 43, 44, 45, 47, 51, 58- 62, 66, 69, 70, 72- 78, 95, 107, 108, 112- 114, 120-124, 128, 129, 134, 136- 138, 143, 145- 147, 149, 150- 152, 157, 159-163, 173- 175, 179-181, 184, 187-199, 203, 204, 208-211, 213, 215, 219, 220, 223- 227, 237, 258, 259, 265, 285, 286, 291, 292, 294
Interface vii, ix, 1, 2, 6, 10, 14, 16, 21, 22, 24, 25, 31-34, 36, 38-40, 43, 45, 47, 49, 50, 51, 53- 58, 60-66, 68, 72, 74, 76, 80, 82, 97, 99, 101, 104, 105, 108-110, 113, 114, 120-124, 126, 128, 129, 131, 134, 135, 147, 149, 153, 154, 159, 161-163, 173-175, 183, 184, 187-190, 195-207, 209, 212, 215, 222, 223, 225, 226, 232, 239, 242, 262, 280
IP Industry 4, 6, 11

Japan 5, 132

Keating 4, 7, 18, 19, 66
Korea 7

Lithography Scaling 255

Meta-Methods 21, 22, 41
Microcontroller 27, 104, 159, 161, 163, 164, 168
Microprocessor 57, 123, 124, 142, 147, 159, 161, 162, 165, 172, 177, 178, 184, 188, 257, 259

MIPS 71, 73, 86, 88, 147
Modelling 15, 25, 27, 187, 191, 198, 199, 200, 201, 203, 207, 208, 210, 212, 225, 226, 233, 258, 296
Morphics 10
Motorola 8, 11, 61

Network-on-Chip 53, 260
Nexperia 15, 33, 67-74, 76-96, 143
Nios 159, 162, 164, 165, 168, 174, 176, 178, 179, 182, 185

OAK 10
OCPIP 108, 118
OMAP i, ix, 15, 97, 98, 99, 100, 102, 103, 104, 105, 106, 107, 117, 143
On-Chip Bus (OCB) vii, ix, 5, 6, 10, 11, 14, 15, 30-33, 35, 36, 43, 50, 53, 57, 59, 73, 74, 75, 85- 87, 103, 108, 112, 120-122, 125, 126, 128, 129, 131, 133, 134, 135, 147, 150, 151, 154, 159, 161, 162, 164, 174, 177, 178, 180, 185, 188, 189, 195, 197-199, 201, 203-209, 216, 217, 218, 221, 222, 226, 260
Open Core Protocol (OCP) 108, 111, 112, 113, 117, 226
Open Mobile Application Processor Interface (OMAPI) 101, 118
Open SystemC International 188, 195, 200, 226
Optical Proximity Correction (OPC) 63, 65, 287, 288, 289

Performance Scaling 255, 256, 257, 258, 260
Peripheral Component Interface (PCI) 53, 60, 72, 109, 110, 113, 123, 131, 137, 162
Philips ix, 11, 15, 24, 26, 27, 28, 30, 33, 67, 68, 69, 71, 80, 85, 95, 143
Platform FPGA ix, 15, 16, 27, 141, 142, 144, 150, 151, 152, 157
Platform-Based Design vii, viii, ix, x, 1, 6, 11-15, 18, 19, 21-30, 32, 33, 37-39, 43, 44, 67-69, 71, 80, 83, 84, 95, 97-103, 107, 108, 114-119, 120, 128-134, 138, 139, 141-144, 149, 151, 152, 153, 156, 157, 160, 187-200, 203, 205, 206, 207, 210-216, 219, 220, 221, 222, 223, 224- 226, 240, 241, 291, 292
Power Dissipation 34, 255, 256, 261, 262, 263, 264, 274
PowerPC 10, 120- 123, 128, 129, 133, 140, 146, 147, 149, 150, 151, 154, 155, 157, 162, 295
PrimeXsys Platforms 15, 187, 189, 190, 193, 197, 214, 219, 220, 224- 226
Programmable Logic 159, 160
Programmable Logic Device 159

Rapid Prototyping x, 160, 223, 229, 239, 243, 251
Register-Transfer Level (RTL) vii, 4, 5, 6, 7, 24, 26, 30, 32, 33, 53, 57, 59, 61, 64- 66, 103, 123, 134, 136, 188, 191, 201, 205, 207, 209, 213, 217, 223, 232, 233, 259, 271
Reliability 3, 28, 34, 36, 255, 256, 262, 279, 282
Resolution Enhancement Technique (RET) 287, 289
Reuse vii, viii, 1, 3-13, 18, 19, 23, 30, 32, 35, 37, 41, 43, 44, 53, 61, 66, 71, 74, 95, 98, 101, 107, 116, 117, 120, 122, 123, 128, 137-139, 141-145, 149, 150, 151, 153, 157-159, 190, 195, 213, 214, 218, 222, 224, 226, 258
Reuse Methodology Manual 4, 6, 7, 19
Revolution vii, viii, 1, 9, 67, 142, 150

Scotland iii, 4, 7, 8
Scottish Enterprise 8

Self-Reconfiguring Platform 141, 142, 154- 156
Signal Integrity 6, 36, 59, 255, 256, 269, 276, 289, 292
SoC Revolution vii, viii, 3, 4, 7, 9, 22, 27, 29, 30, 33, 35, 44, 66, 157
SOCWorks 118
ST Microelectronics (ST) 10, 11, 101, 250
Stack 15, 52, 80, 82, 151, 155, 174, 175, 179, 181, 188, 189, 293
Standards iii, viii, 4, 5, 10, 11, 23, 24, 35, 38, 40, 51, 59, 69, 81, 109, 113, 120, 122, 123, 124,
    127, 136, 138, 139, 146, 187, 188, 195-198, 201, 219, 220, 222
Star-IP 187
Surviving the SOC Revolution iii, iv, viii, 2, 12, 19, 44, 118, 157
Sweden 7
System Design iii, vii, 5, 6, 17, 21, 22, 25- 27, 38, 39, 40, 45, 47, 48, 52-60, 69, 75, 97, 118,
    141, 150, 157, 158, 160, 164, 184, 185, 198, 200, 203, 206, 207, 226, 229, 230, 231, 235,
    236, 243, 251, 290, 294
System on a Programmable Chip (SOPC) ii, ix, 15, 16, 159, 161- 163, 174, 180, 184, 185
SystemC iii, 45, 57, 198, 200, 202, 208, 209, 214, 215, 226
System-in-Package (SiP) 10, 39, 293
System-Level Integration 7, 8, 11, 53, 140
System-on-Chip (SoC) i, ii, iii, vii- x, 1, 2, 4, 6-13, 15, 17, 18, 19, 21, 22, 23, 26-35, 37-39,
    41, 44, 45, 47-51, 53- 55, 56- 58, 60- 63, 66, 67, 69, 71-73, 75, 76, 77, 95- 98, 100- 102,
    107- 109, 111, 112, 114- 117, 119- 121, 123, 124, 127- 129, 131, 133, 136- 140, 142, 143,
    150- 152, 157, 160, 187-191, 193, 197- 203, 205, 206, 210, 213, 214, 216, 219, 220, 222-
    224, 226, 227, 229- 232, 239, 255- 257, 259, 260- 266, 268, 269, 272, 275, 276, 278- 287,
    289, 290-295
SystemVerilog 57

Taiwan 7
Technology Integration Challenge 255
Tensilica 10, 16
Texas Instruments (TI) i, ix, 11, 15, 24, 38, 39, 61, 97, 99, 101, 102, 104, 105, 107, 108, 112,
    114, 116, 117, 118, 143
Toshiba 5, 11
Transaction 30, 52, 54, 73, 74, 75, 76, 86- 90, 103, 121, 122, 135, 198- 201, 208, 209, 215
Transaction-Based Verification 30
Transaction-Level Modelling 208

Variability 34, 255, 256, 276, 277, 278, 279
Verification ix, 4, 6, 7, 13, 19, 21, 22, 24, 29, 30- 33, 36, 38- 40, 45, 48, 53-59, 103, 108, 111,
    113-115, 119- 122, 124-140, 152, 187- 190, 193, 195- 198, 200, 203, 205, 207, 209, 213,
    214, 216- 218, 220, 223- 227, 229, 230, 231, 243, 247, 251, 290, 292
Verification and Validation 187, 188, 190, 195, 213, 214, 223, 224, 226
Verilog 51, 52, 126, 160, 181, 183, 196, 203, 215
VHDL 160, 196, 203, 237, 244, 245
Video 33, 67, 68, 69, 71, 95, 96
Virtual Component (VC) 6, 8, 12, 19, 44, 59, 108, 196, 229, 231, 234, 235, 244, 245, 251
Virtual Component Exchange 8, 11
Virtual Component Interface (VCI) 6, 74, 108
Virtual Socket 1, 4, 5, 6, 18, 35, 45, 51, 66, 108, 118, 123, 188

Virtual Socket Interface Alliance (VSIA) iii, ix, 1, 4, 5, 6, 11, 12, 18, 19, 35, 45, 51, 58, 59, 61, 66, 108, 118, 123, 138, 188, 195, 196, 198, 226

Xilinx ix, 15, 16, 23, 27, 29, 141-144, 147, 149, 150, 153, 156, 157, 237, 238, 241, 242, 244, 248, 252

Yield 65, 125, 255, 256, 261, 279, 283- 286, 293, 296